SpringerBriefs in Modern Perspectives on Disability Research

This book series on disability research is a comprehensive collection of research on disability and related issues. The series is designed to promote interdisciplinary collaboration and exchange, bringing together scholars and practitioners from different fields to share their perspectives and insights. Disability research is an interdisciplinary field that examines the social, cultural, historical, and political dimensions of disability. It encompasses a wide range of topics, including disability rights, accessibility, assistive technologies, healthcare, education, employment, and social welfare. Disability research scholars employ a range of theoretical and methodological approaches to understand the experiences of people with disabilities, as well as the ways in which disability intersects with other social identities such as race, gender, sexuality, and class.

The series seeks to advance knowledge and understanding of disability by publishing rigorous, innovative, and relevant research. It aims to promote disability rights and social justice by highlighting the ways in which people with disabilities are marginalized and discriminated against in society, and advocating for greater social inclusion and accessibility. The series also seeks to inform policy and practice by disseminating research findings that can help to shape policy decisions and contribute to positive social change.

Ajay Kumar
Editor

Integrating Disability Care in Major Depressive Disorder Through Neurological and Mental Health Perspectives for Empowerment

Editor
Ajay Kumar
Department of Biotechnology, Faculty of Engineering and Technology
Rama University
Kanpur, Uttar Pradesh, India

ISSN 3004-9709 ISSN 3004-9717 (electronic)
SpringerBriefs in Modern Perspectives on Disability Research
ISBN 978-981-95-8954-8 ISBN 978-981-95-8955-5 (eBook)
https://doi.org/10.1007/978-981-95-8955-5

This Springer imprint is published by the registered company Springer Nature Singapore Pte Ltd.
The registered company address is: 152 Beach Road, #21-01/04 Gateway East, Singapore 189721, Singapore

Foreword

Major Depressive Disorder (MDD) is one of the leading causes of disability worldwide; however, existing mental healthcare frameworks often fall short in adequately addressing its disabling consequences. *Integrating Disability Care in Major Depressive Disorder Through Neurological and Mental Health Perspectives for Empowerment* emerges at a critical time, as the limitations of fragmented treatment approaches become increasingly apparent and global concerns surrounding mental healthcare continue to intensify. A defining strength of this book lies in its holistic and multidimensional perspective. By framing depression as a pressing priority for academic dialogue and clinical advancement, it investigates the condition through four interrelated lenses vulnerability, emotional suffering, social exclusion, and systemic barriers to care. This work goes beyond traditional clinical concerns for its scope, management, and symptomatology by seeking to view its subject from a functional perspective an approach that forms an integral part of promoting patient empowerment. Building upon significant research undertaken within neurological sciences, mental health, and disablement studies, the book identifies many key themes that are pertinent and important for today's academic and clinical requirements. Equally important is the attention it pays to the realities of healthcare today, stressing the importance of the balance between the three elements of human-centered approach, technological help, and teamwork. A strong focus on capacity building, education, and professional competencies, which are all highlighted throughout the book, makes it unique and important within the field.

I congratulate Prof. Ajay Kumar and the contributing authors for their scholarly dedication and commendable efforts. I am confident that this book will make a meaningful contribution toward transforming disability-informed mental health care.

Dr. Syed Muhammad Ali
Assistant Professor at Weill-Cornell
Medicine Qatar
Hamad Medical Corporation
Doha, Qatar

Preface

Major Depressive Disorder (MDD) affects not only emotional well-being but also cognitive, social, and functional aspects of life, making it one of the leading causes of disability worldwide. Despite significant progress in diagnosis and treatment, persistent disability remains a major challenge for many individuals living with the disorder. Recognizing the need for comprehensive disability support alongside clinical treatment inspired the development of this book. The goal of the book *Integrating Disability Care in Major Depressive Disorder Through Neurological and Mental Health Perspectives for Empowerment* is to bridge the knowledge gaps between disability education, mental health, and neuroscience. The three aspects are treated independently in current methodologies, which results to provide a comprehensive model of treatment that draws upon all three concepts Therefore, the goal of the book is to offer a thorough treatment model that incorporates all three ideas. The book is structured to connect every theory to real-world applications. The idea of depressive impairment and the associated psychological theories are emphasized in the book's opening portions. However, systems medicine, information technology, and teamwork are covered in the following sections. However, the majority of the book's early sections concentrate on empowerment models that encourage depressed people to take control of their life. The book also discusses clinical practitioners' education and ability to handle the disorder's intricacies. As a result, the book addresses curriculum development, competency, and disability evaluation. A key focus of the book is strengthening the educational capacity of clinical practitioners so they are better equipped to address the complexities associated with Major Depressive Disorder (MDD). It also examines broader structural challenges related to depression, including the long-term mental health impact of the pandemic and ongoing limitations in resources and policy frameworks. By addressing these critical issues, the book aims to support the advancement of mental health systems at both professional and policy levels. Designed for graduate students, healthcare professionals, researchers, and advocates in mental health and disability studies, this book seeks to stimulate meaningful dialogue, enhance understanding, and promote positive change. Above all, it emphasizes that individuals with Major Depressive Disorder deserve care that recognizes their abilities, preserves their dignity, and empowers them to participate fully and actively within their communities.

Kanpur, India Ajay Kumar

Acknowledgements

I would like to express my sincere gratitude and deep appreciation to Dr. Pranav Singh, Director, Rama University, Kanpur, India, for his outstanding support, encouragement, and motivation, which greatly contributed to the successful completion of this book. I am also deeply grateful to Dr. R. K. Yadav, Controller of Examinations, Rama University, Kanpur, for his exceptional personal and professional support and constant inspiration throughout this endeavor.

I am delighted to thank all the contributing authors for their excellent scholarly contributions, which have enriched the quality and depth of this book. I also extend my sincere thanks to Gabriel Bennett and Emma Goodall (Series Editors, *SpringerBriefs in Modern Perspectives on Disability Research*), Ms. Jacqueline Eu (Editor, Behavioral & Health Sciences), and Mr. Rajesh Janakiraman (Production Coordinator), Springer, for their professional guidance and efficient management of this project.

I greatly appreciate the support and encouragement of my students Dr. Fariya Khan, Dr. Indrajeet Singh, and Ms. Laxmi Singh, whose insightful discussions and constructive comments helped shape this book. I am also thankful to Dr. Vijai Singh, Dr. Rakesh Kumar, and many others whose names may not be mentioned here but who have directly or indirectly contributed to the development of this work.

I am also thankful to Dr. Javed Iqbal, Hamad Medical Corporation, Doha, Qatar, for his valuable support, guidance, and encouragement during the course of this work. His insights and cooperation are gratefully acknowledged.

My deepest gratitude goes to my beloved wife, Shraddha Singh, for her unwavering support, patience, and constant inspiration. I extend my love and affection to my children, Nishit and Anshika, who missed me during the course of this project. I also seek the blessings of my elder brothers, Er. Prem Chandra and Dr. Ram Kumar, who have always been guiding figures in my personal and professional journey.

I warmly thank the faculty and staff of Rama University for providing a supportive and stimulating working environment. I am aware that despite our best efforts, the first edition may contain inadvertent errors, and I would sincerely welcome feedback from readers to further improve future editions of this book.

I am profoundly thankful to the Almighty God for granting me the wisdom and strength to complete this work. Finally, I dedicate this book to my beloved parents, whose countless blessings and unwavering faith made this achievement possible.

About This Book

Integrating Disability Care in Major Depressive Disorder Through Neurological and Mental Health Perspectives for Empowerment presents an extremely pertinent and comprehensive analysis of Major Depressive Disorder as a primary factor of disability in the worldwide domain. The status of disability in Major Depressive Disorder has attracted correspondingly lesser attention in the healthcare processes being developed in the healthcare systems globally. Depression has been extensively studied from both clinical and psychiatric points of view. However, the book presents a highly integrated conceptualization of the topic by correlating neurological and mental health perspectives of disability care, in a manner particularly suited to the post-COVID era, during which the global relevance of mental health challenges has increased significantly. At a time when the prevalence of depressive impairment is rising substantially, the book emphasizes person-centered approaches grounded in empowerment and active patient participation in healthcare processes. It advocates shifting individuals from passive recipients of care to collaborative partners in mental health management an approach that is particularly relevant in contemporary practice and in the post-pandemic landscape of evolving global healthcare needs. Additionally, the book provides a comprehensive examination of neurobiological dysfunction within prefronto-limbic circuits and its association with psychosocial impairment and long-term disability among affected individuals.

Contents

About the Editor

Dr. Ajay Kumar is a Professor of Biotechnology and currently serves as Dean (Research and Development) and Dean (Engineering) at Rama University, Uttar Pradesh, Kanpur, India. He has over 21 years of extensive experience in teaching, research, and academic administration. Dr. Kumar has demonstrated proven expertise in Genomics and Proteomics, Bioprocess Engineering, Bioinformatics, Microbiology, Industrial Microbiology, Genetic Engineering, Environmental Biotechnology, Fermentation Technology, and Food Biotechnology. He has held several key academic and administrative positions in well-renowned universities and engineering institutions. He earned his M.Tech. in Biotechnology from the Institute of Engineering & Technology, Lucknow, India, and completed his Ph.D. at the ICAR–Central Institute for Research on Goats (CIRG), Mathura, India. His core research interests include computational vaccine and drug development, cancer biology (genomics and proteomics), and fermentation technology. Dr. Kumar has an impressive research portfolio, with 110 research articles, 50 book chapters, 6 books, and 7 patents to his credit. He has also served as Chief Editor, Associate Editor, Editorial Board Member, and Reviewer for several reputed peer-reviewed international journals. As a recognized research guide, he has successfully supervised M.Tech. and Ph.D. scholars, many of whom have been awarded their degrees, and he continues to mentor doctoral and postgraduate researchers. In addition, he is a member of the Board of Studies and Academic Council of Rama University, Kanpur, as well as the Regional Food Research and Analysis Center, Department of Horticulture and Food Processing, Lucknow, Uttar Pradesh. Dr. Kumar is an active member of professional bodies, including the International Association of Engineers (IAENG) and INSA, and has provided consultancy services in vaccine research, contributing significantly to academia and industry.

Chapter 1
Introduction to Disability Care in Major Depressive Disorder

Ajay Kumar, Juveriya Israr, and Shabroz Alam

Abstract Major depressive disorder (MDD) is a prevalent mental condition that significantly impacts everyday functioning. This chapter discusses disability care in Major Depressive Disorder (MDD), with a focus on the bio-psycho-social aspects that impact deficits in emotional regulation, cognition, movement, work performance, and social interaction. Disability theories from the medical, social, and recovery-based spheres are examined, along with diagnostic features and functional limitations. In addition to discussing culturally safe, trauma-informed, and person-centered treatment, this article delves into concrete strategies to assist with day-to-day tasks, psychological and social rehabilitation, workplace engagement, and emergency preparedness. Caretakers, rights-based methods, anti-stigma, and interdisciplinary collaboration are major themes throughout the chapter. It is believed that digital and assistive technology may help people become more independent, more integrated into their communities, and more accessible. The chapter emphasizes that providing high-quality, comprehensive assistance for MDD includes caring for individuals with impairments.

Keywords Major Depressive Disorder · Disability care · Functional impairment · Psychosocial rehabilitation · Vocational support · Activities of daily living

A. Kumar (✉)
Department of Biotechnology, Rama University, Kanpur, India
e-mail: ajaymtech@gmail.com

J. Israr
Institute of Biosciences and Technology, Shri Ramswaroop Memorial University (SRMU), Barabanki, India

J. Israr · S. Alam
Department of Biotechnology, Era University, Lucknow, India

A. Kumar (ed.), *Integrating Disability Care in Major Depressive Disorder Through Neurological and Mental Health Perspectives for Empowerment*,
SpringerBriefs in Modern Perspectives on Disability Research,
https://doi.org/10.1007/978-981-95-8955-5_1

1.1 Introduction

Millions of people around the world have Major Depressive Disorder (MDD), which is one of the main causes of disability at all ages. MDD is known for causing emotional and mental problems, but it can also affect other types of mood disorders. Deterioration in focus, memory, drive, stamina, and social interaction are only a few of the many ways in which depression impacts daily functioning. According to (Chakrabarty et al., 2016), this condition makes it more challenging to get clinical treatment, long-term care, rehabilitation, and community participation. Care for people with disabilities explains and deals with the practical effects of MDD. Rather than focusing only on treating symptoms, this approach takes into account how depression affects a person's social life, professional life, and capacity to take care of themselves. Autonomy, inclusion, and important life roles are promoted in disability care via mental health, disability studies, and rehabilitation. Inadequate motivation or cognitive slowness are examples of internal causes of disability; stigma, inaccessibility, and a lack of support are examples of external causes (Dixon et al., 2001). Providing assistance to those with impairments has grown in significance as a means of combating depression in recent times. Because medication and psychotherapy aren't always enough to fix functional deficiencies, it's clear that we need broader initiatives to boost autonomy, quality of life, and daily functioning. The mental health community has recognized, thanks to the disability rights and human rights movements, the need of respect, empowerment, and person-centered treatment for rehabilitation (Ee et al., 2020). Caring for persons with impairments caused by MDD is discussed in this chapter along with related concepts, models, and methods. Functional ability, restrictions, and day-to-day support for people with the disease are some of the issues covered. The goal of this chapter is to help students, doctors, support staff, and caregivers understand depression from a bio-psycho-social and recovery-focused perspective so they may empower depressed persons to take charge of their own lives, overcome barriers, and become more actively involved in their treatment. Restoring agency, reducing suffering, and facilitating meaningful community participation are all tenets of disability care.

1.2 Understanding Major Depressive Disorder

MDD is a complicated mental health disease that is characterized by long-term problems with mood, thinking, and physical health. It is one of the most common mental illnesses and has an effect on work, school, relationships, and quality of life. Even though MDD is mostly about emotions, it can cause a lot of problems in daily life. Because MDD often leads to disability and functional impairment, comprehending it requires an analysis of both its psychological and functional dimensions. MDD can be very different from person to person, even though it happens in episodes. Some people only have one short episode, while others have chronic depression. Depression is caused and kept going by biological vulnerability, trauma, stress, social factors, and personal history. Because of this diversity, disability care needs to be different for each person based on their needs, how they present, and their life situation (Woo et al., 2016).

1.2.1 Diagnostic Features

The Diagnostic and Statistical Manual of Mental Disorders, Fifth Edition (DSM-5) outlines the precise criteria for diagnosing MDD. For a diagnosis, there must be at least five symptoms that show a change from how things used to work and that have been there for at least two weeks. You need to have either low mood or anhedonia, which is not being interested in or enjoying things.

1.2.1.1 Core Symptoms Include

The symptoms of MDD include an overwhelming sense of melancholy, a severe decline in interest in and pleasure from most things, and extreme changes in appetite and weight, which may be either increased or decreased. Individuals may also experience difficulties with sleeping, including hypersomnia or insomnia, psychomotor agitation or retardation, extreme guilt or feelings of worthlessness, impaired concentration, decision-making abilities, and thoughts of suicide or death, as well as attempts to end their own lives. All of these symptoms need to make it hard for you to operate in important areas of your life, such as social interactions or at work. One must comprehend the link between mental symptoms and functional impairment in order to comprehend MDD, which is more than just an emotional disease; it hinders everyday functioning, meaningful relationships, and involvement in profession, school, and community life. For a diagnosis to be determined, it is necessary to rule out the possibility that these symptoms are caused by any medicine or medical condition. This diagnosis does not include bipolar disorder, thyroid problems, or mood disorders brought on by drugs. These factors emphasize how common MDD is and how important it is to have disability-aware treatment that addresses pain and functional limits.

1.2.2 Course and Variability

There is a wide range in the onset, persistence, and severity of MDD symptoms, as well as their influence on everyday functioning. Many subtypes of MDD might emerge with the passage of time. Knowing diversity is essential for developing high-quality disability care and tailoring treatment to each individual's needs (Gustavson et al., 2018).

1.2.2.1 Episodic and Time-Limited Presentations

A single episode of MDD may occur and either go away on its own or be helped by therapy. These episodes might happen as a consequence of unexpected emotional or psychological stress, major life changes, or loss. Even though many people's functioning abilities are severely impaired during an episode, they could recover to how they were before. This strategy highlights the need of prompt medical attention in preventing permanent impairment.

1.2.2.2 Recurrent Depression

A significant proportion of individual's experience recurrent episodes characterized by alternating periods of depression and remission. Recurrence is more likely in cases of early onset, incomplete recovery, persistent psychosocial stress, or a family history of mood disorders. Persistent monitoring, developing relapse prevention techniques, and providing long-term disability support should be prioritized since recurring episodes may lead to cumulative functional impairment (Solomon et al., 2000).

1.2.2.3 Chronic and Persistent Forms

Some individuals may experience persistent or intermittent symptoms of depression that never fully resolve. Chronic MDD exacerbates social, occupational, and cognitive functioning when coupled with anxiety, a history of trauma, or other medical conditions. (Shelton & Hollon, 2012) say that when symptoms don't go away, Integrated care approaches could be required. Assistance for persons with impairments, psychosocial rehabilitation, and clinical therapy are all examples of such approaches.

1.2.2.4 Treatment-Resistant Depression

Treatment-resistant depression (TRD) is when someone doesn't get better after trying at least two scientifically proven treatments for depression at the right dose and time. This condition affects some people who are depressed. TRD may worsen disabilities, healthcare utilization, and other biopsychosocial factors. Individuals with TRD may find ketamine-based therapies, TMS, ECT, or significant psychological support beneficial (Gaynes et al., 2020).

1.2.2.5 Cultural, Social, and Biological Influences

Multiple cultural, social, and biological variables impact the onset and course of MDD. Depression is more often associated with mental or emotional issues in certain cultures than with physical symptoms in others. Prolonged bouts or recurrence might be caused by poverty, discrimination, trauma, and insufficient healthcare. Symptoms and the effectiveness of therapy are impacted by a myriad of biological variables, including genetics, neurochemical imbalances, medical comorbidities, hormonal changes, and more (Kessler & Bromet, 2013).

1.2.2.6 Implications for Disability Care

There is no set course for MDD, therefore disability treatment must be adaptable and based on each person's unique symptoms, situation, and functional goals. Because depression affects people in so many different ways, it can't be treated with a single type of care. Therefore, in order to offer quality care to individuals with disabilities, it is crucial to conduct comprehensive assessments, develop individualized plans, and maintain open communication channels among patients, service providers, and healthcare professionals. This way, the supports may be adjusted as the condition progresses.

1.3 Functional Impairment and Disability in MDD

MDD affects more than just mood. The symptoms of this condition can make it very hard for a person to think clearly, do everyday tasks, keep up with their relationships, and take on important responsibilities. Disability, defined as these functional restrictions that often persist even when mood symptoms subside, is an important part of the disorder's clinical picture. It is important to determine the severity and kind of functional impairment in order to create effective disability care programs. Physical, social, occupational, emotional, and cognitive functioning are the five primary domains often impacted by MDD (Fig. 1.1). Impaired motivation, for example, might worsen impaired occupational performance, for example, because of the dynamic interaction between these domains (IsHak et al., 2016).

1.3.1 The Concept of Functional Disability

Functional disability, resulting from the cognitive, behavioral, and physical manifestations of MDD, alongside emotional symptoms, refers to the incapacity to execute daily tasks, activities, and responsibilities essential for autonomous living and significant engagement (IsHak et al., 2016). People with MDD often feel very sad, anxious, angry, or hopeless, which can make it hard for them to make decisions, start tasks, and interact with others (Sheppes et al., 2015). Academic success, job duties, and everyday organization may all take a hit when processing speed, memory, attention, and executive functioning are impaired. One way to isolate oneself and put pressure on relationships is to lose interest in social activities, avoid connecting with

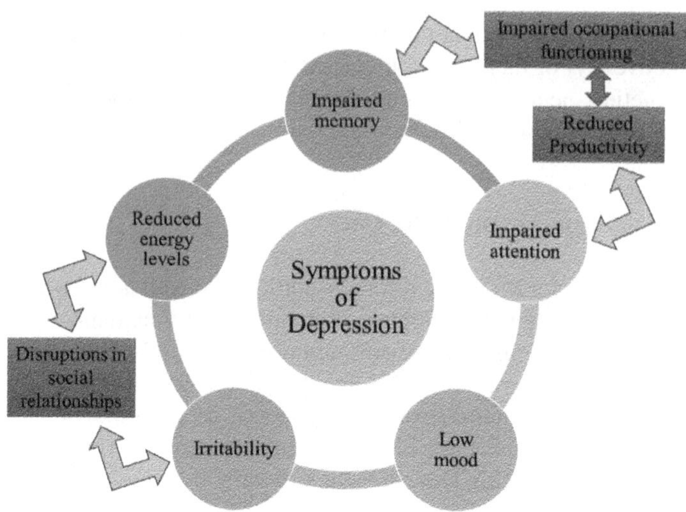

Fig. 1.1 Domains of functional impairment in major depressive disorder

others, and communicate less (Rubin et al., 2009). Problems with self-care, job, and everyday life may arise from a lack of energy, motivation, and the ability to begin tasks. MDD can make you tired and uncomfortable, which can slow down your movement, stamina, and physical or work-related activities (Penninx et al., 2013). MDD therapy that focuses on disabilities is necessary because these emotional, cognitive, social, and physical impairments have a negative impact on quality of life (Hamilton et al., 2012).

1.3.2 Cognitive Impairment

Disregard for the cognitive impairment caused by MDD is common. Cognitive markers reliably predict functional impairment better than mood intensity alone, despite the fact that emotional symptoms get the majority of therapeutic attention. Cognitive impairments are a major target for disability-oriented treatment since they may appear during severe depression, continue after partial remission, and continue even after mood symptoms have entirely gone (Gonda et al., 2015). People with MDD often talk about psychomotor slowness, which slows down brain processes and makes it hard to react, do things, and talk to people (Chakrabarty et al., 2016). Executive functioning, which includes planning, organizing, setting priorities, and starting tasks, is often impaired, making it hard to do everyday tasks. Problems with attention and working memory can make it hard to stay focused, ignore distractions, and remember things for later use. Work, education, and daily life may all be affected by this (Mierau, 2025). Cognitive rigidity, excessive rumination, and poor problem-solving abilities may also make it hard to make decisions and respond adaptively to problems. Academic performance can suffer due to difficulties focusing, processing information slowly, and not being well-organized (Polderman et al., 2010); occupational performance can suffer due to inefficiency, forgetfulness, and problems with complex task management (Salvagioni et al., 2017), and social interactions can be challenging. People with depression can greatly benefit from cognitive remediation, structured routines, compensatory aids, task modification, and accommodations at school or work. It is crucial to recognize cognitive impairment as a key component of MDD in order to provide good disability care (Woo et al., 2016).

1.3.3 Occupational Impairment

Occupational functioning is greatly affected by MDD, which affects productivity, attendance, job performance, and interpersonal interactions. It is a prominent cause of workplace impairment globally. Depression may have a lasting impact on one's ability to work, which can start early on, continue even when the disease is in partial remission, and even after emotional symptoms have subsided (Lagerveld et al., 2010). Patients with MDD often experience absenteeism, presenteeism, and difficulties with task initiation and organization due to diminished motivation (Evans-Lacko & Knapp, 2016). These worries could lead to losing a job, being demoted, not moving up in your career, or retiring. Disability care must address vocational impairment since employment is vital to financial stability, identity, and social participation. The following are examples of reasonable accommodations that can help individuals with disabilities in the workplace: reduced or flexible hours, different tasks or workloads, increased supervision or mentoring, assistance with task organization and prioritization, low-stimulation or quiet work environments, a gradual return to work after leave, and collaboration with occupational therapists. A collaborative strategy among employers, clinicians, disability support professionals, and the individual is necessary to create a supportive workplace that recognizes the variable nature of MDD. This strategy can enhance an individual's capacity to sustain employment, mitigate functional impairment, and elevate overall well-being (McDowell & Fossey, 2015).

1.3.4 Social and Interpersonal Disability

Dissimilarities in self-perception, social-information processing, and interpersonal functioning characterize MDD. Disabling effects of interpersonal issues may last even after a person's mood improves. A lack of social connection, which is critical for mental health, might reduce quality of life and make recovery more difficult (Hames et al., 2013). Patients with MDD frequently avoid social gatherings and communal living due to fatigue, diminished interest in previously enjoyed activities, and heightened self-consciousness (Kim et al., 2025). Depression can also damage friendships, family, and close relationships because it makes people emotionally unavailable, hard to talk to, and irritable (Hames et al., 2013). People don't get help, admit their problems, or fully participate in social situations because they are afraid of being criticized or because they have internalized self-stigma (Link et al., 2001). According to (Vos et al., 2025), people with cognitive impairments and negative thinking tend to misread social signals, which might cause them to misjudge the intentions or feelings of those around them. Some people may experience increased sensitivity, impatience, and a decreased capacity to tolerate frustration as a result of trying to avoid touch as a means of reducing stress or shame (Lieberz et al., 2021). Isolation and withdrawal may become a vicious cycle when depression is accompanied by problems in interpersonal connections (Wang et al., 2018). They may also heighten feelings of loneliness, make it harder to ask for help from loved ones, and make it harder to finish job assignments. Therefore, it is essential that disability care incorporates social functioning. People with disabilities can benefit from structured social skills training, supportive counseling or psychotherapy that focuses on interpersonal patterns, peer support programs, family education that improves communication and understanding, community engagement initiatives, and meaningful participation in social environments. These programs can aid in long-term recovery and improve quality of life.

1.3.5 Physical and Somatic Disability

Although MDD is most often associated with mental or emotional health issues, it may also manifest with severe physical and somatic symptoms that impair daily functioning and may persist long after mood symptoms resolve. According to (Berk et al., 2023), this is the reason why these symptoms must be recognized in order to get complete disability treatment. In addition to worsening depressive symptoms, limiting mobility, and diminishing social and vocational participation, persistent pain (including headaches, musculoskeletal pain, gastrointestinal complaints, or widespread body aches) may be a result of MDD. Depressed people often have poor endurance and chronic fatigue, making it difficult for them to engage in regular physical activity, go to work, and do ordinary chores. According to (Ghanean et al., 2018), this may cause significant functional impairment in some individuals. Fatigue, cognitive impairments, emotional instability, and impairment may be worsened by the common sleep disorders that affect many people (Harvey, 2011). Task performance, social engagement, and self-care might be impacted by psychomotor alterations such as slower movements, speech, or reactions, along with restlessness and agitation (Sobin, 1997). The inability to take care of oneself, complete household chores, or participate in community activities is worsened by physical pain and exhaustion, and by the worsening of mood as a result of depression (Berk et al., 2023). Interventions that take disability into account can help with these physical aspects in the following ways: improving strength and endurance through graded exercise or physiotherapy; managing pain through pharmaceutical and non-pharmacologic means; conserving energy through activity pacing and energy conservation; and addressing both psychological and physical symptoms through integrated care models. Rehabilitation, functional capacity, and quality of life may all be improved with disability care that addresses the mental, emotional, and physical components of depression.

1.4 Models of Disability Applied to MDD

In order to diagnose functional deficiencies and guide evaluation, therapy, and support, conceptual frameworks, also known as models of disability, are required for MDD. In order to better assist individuals with disabilities, physicians, caregivers, and lawmakers might use disability models to guide their decision-making. The three most common models are medical, social, and biopsychosocial/recovery-oriented. Care is impacted by each perspective (Chow et al., 2022).

1.4.1 The Medical Model

Functional issues such as cognitive slowing, lack of motivation, and trouble sleeping are attributed to MDD in the medical model of disability. The treatments in this paradigm focus on making an accurate diagnosis, relieving symptoms, and providing therapeutic therapy (Chow et al., 2022). Psychotherapy, including evidence-based approaches such as cognitive-behavioral therapy (CBT), interpersonal therapy (IPT), and psychodynamic therapy, and pharmacotherapy, including SSRIs, SNRIs, and other antidepressants, are common medicinal model therapies for MDD (Kennedy et al., 2016). While the medical paradigm is effective in alleviating major depressive disorder symptoms, it falls short when it comes to assisting those with impairments. Despite the fact that things like stigma, locations that are difficult to reach, and stress in the workplace and one's family may all make things harder to achieve, people often disregard social and environmental variables (Kirkbride et al., 2024). Individuals may nevertheless have substantial constraints in their ability to participate in occupational, educational, or social activities even after their symptoms have resolved. According to (Bottema-Beutel et al., 2020), if we confine our understanding of disability to illness, we run the danger of reductionism, which erodes human agency and lived experience by painting functional limitations as flaws instead of impediments that need adaptive therapies. We can learn more about the kind of impairment in MDD if we integrate medical model research with social work and rehabilitation-based techniques. For instance, according to (Winstein et al., 2016), antidepressants may help with symptoms like low mood and exhaustion, but a more holistic approach might include changing one's surroundings, learning social skills, and getting structured vocational aid.

1.4.2 The Social Model

The social model challenges the common wisdom that illness is the root cause of disability and instead attributes it to societal and environmental factors. This theory states that accessibility, social attitudes, and structural factors define the severity of disability, whereas symptoms and external factors dictate the functional constraints (Oliver, 2013). MDD symptoms may be exacerbated by social and environmental factors. People with mental illness may already feel stigmatized, alone, and unworthy, and prejudice and discrimination against them may make them feel much worse. Because of real and perceived stigma, people may be hesitant to ask for help or give their all in their social, academic, and work responsibilities (Hinshaw & Stier, 2008). Problems at work, such as long hours, difficult jobs, and bosses who don't understand people with MDD, make it hard for them to hold down a job or move up the corporate ladder (McDowell & Fossey, 2015). Socioeconomic disparities, geographical barriers, and an insufficient healthcare system may all pose challenges to mental health treatment, rehabilitation, and psychosocial support (Mongelli et al., 2020). Environmental and social constraints may impede daily functioning and meaningful engagement. Things like a lack of social support, limited community resources, and unsuitable public spaces are examples of these obstacles. (Kirkbride et al., 2024) states that the social model of disability care prioritizes supporting individuals with impairments, integrating them into society, and accommodating their requirements. Policies that encourage fair access to healthcare and social services, employment adjustments, flexible scheduling, anti-stigma campaigns, community-based peer support, and education for families or caregivers to assist overcome relational obstacles are all part of this.

1.4.3 The Bio-Psycho-Social Model

Disability and MDD are combined in the BPS technique. The BPS approach differs from medical or social therapies in that it places an emphasis on the interplay between biological, psychological, and social elements as the root cause of depression and its functional limits. Effective treatments, according to this paradigm, should target the whole person, not just their symptoms (Chow et al., 2022). This is how it helps with disability care.

1.4.3.1 Biological Vulnerabilities

Biological variables have a role in the onset and maintenance of MDD. Depression may have several causes, including hereditary factors, abnormalities in the serotonergic, dopaminergic, and noradrenergic systems of the nervous system, and changes in hormones such thyroid dysfunction, alterations after giving birth, or dysregulation caused by stress. Depression and performance may be exacerbated by underlying medical issues. Mood regulation, mental acuity, stamina, and motor skills are all impacted by these inherent biological deficiencies. Along with psychological and social variables, they often cause and sustain impairment in MDD (Mehra et al., 2025).

1.4.3.2 Psychological Features

Maladaptive cognitive patterns, such as prolonged negative thinking, rumination, and impaired problem-solving, alongside emotional regulation disorders, including stress sensitivity and ineffective coping mechanisms, characterize MDD. Avoidance, social isolation, and a lack of initiative make it harder to do well in school, at work, and in social situations. These psychological traits may exacerbate disability by disrupting daily functioning and role fulfillment. Cognitive Behavioral Therapy (CBT), stress management, and skills-based therapies may help individuals with depression enhance their functioning and reduce impairment (Beck & Haigh, 2014).

1.4.3.3 Social Determinants

The way depression works is affected by society and the environment. When people in economically and socially unstable areas cannot easily access treatment, rehabilitation, and other necessary resources, their depression symptoms intensify and recovery becomes more challenging. Exposure to trauma or prejudice raises the risk of depression and makes recovery more difficult; social isolation and lack of support from others worsen emotional pain and weaken resilience. Because of prejudice, inadequate support structures, and lack of adaptations, one's professional, academic, and social lives may suffer. Activism, community participation, legislative reforms, and peer support are necessary for individuals with depression to reduce functional impairment, increase inclusion, and recover successfully (Ridley et al., 2020).

1.4.3.4 Implications for Disability Care

Biologic, psychological, and social aspects of managing MDD impairment are all emphasized in the biopsychosocial (BPS) paradigm. Some examples of biological therapies include medication, medical management, and symptom monitoring. Psychological therapies include a wide range of approaches, including talk therapy, cognitive restructuring, and stress management techniques. Efforts to remove stigma, integrate communities, alter educational or employment environments, and engage in lobbying are all examples of social interventions. Functional competency, quality of life, and long-term rehabilitation are all improved with disability-informed care.

Because recovery-oriented methods prioritize independence, purposeful participation, and health, the BPS framework is a vital component of modern MDD disability treatment (Dixon et al., 2001).

1.4.4 The Recovery Model

Rehabilitation changes how we look at functional disability and MDD handicap. Rather than focusing on traditional therapies that try to ease symptoms, recovery fosters personal growth, autonomy, and living a worthwhile life. In doing so, it acknowledges the possibility that residual sadness may enhance independence, social engagement, and life satisfaction (Jacobson & Greenley, 2001).

1.4.4.1 Core Principles of Recovery

Patients with severe depressive illness may reclaim their freedom, find purpose in life, and improve their quality of life via rehabilitation. Restoring autonomy requires reducing reliance and increasing control by giving people the tools they need to make educated decisions about their own care, everyday activities, and life objectives (Carey et al., 2023). Getting back into the things that make you happy, give you a sense of purpose, and strengthen your connections will boost your self-esteem (King & Hicks, 2021). Enhancing quality of life may be achieved via a combination of measures, including symptom reduction, social engagement, occupational engagement, physical health, and emotional resilience strengthening in support of holistic healing.

1.4.4.2 Alignment with Disability Care Principles

People with disabilities are given hope, power, and the freedom to make their own choices as part of the rehabilitation strategy, which is similar to present disability care. Empowerment is giving people the capacity to make decisions, establish goals, and advocate for themselves. The method encourages resilience and hope by demonstrating that full healing is achievable even in the face of ongoing symptoms. Recognizing that there are practical advantages beyond just alleviating symptoms, recovery-oriented therapy prioritizes self-determination, values, and life goals (Mellillo et al., 2025).

1.4.4.3 Practical Implications for Care

The recovery paradigm for MDD disability therapy includes defining individual, clinician, and caregiver goals as well as person-centered planning for social, occupational, and daily living requirements. This method places an emphasis on mentoring, community service, and flexible treatment in addition to an individual's strengths, areas for improvement, aspirations, and social networks. By placing an emphasis on agency, participation, and holistic health, the recovery paradigm helps people with MDD reclaim meaningful functioning and life pleasure (Dixon et al., 2001).

1.5 Principles of Disability Care in MDD

Care for people with MDD disabilities should focus on symptom management, but it should also include functional limitations, social engagement, and quality of life. The requirements and circumstances of each individual are the basis for comprehensive, customized, and recovery-oriented treatment (Marx et al., 2023).

1.5.1 Person-Centered Care

Interventions that take into account disabilities and MDD are rooted on patient-centered treatment. Given that functional constraints and disability experiences are context dependent, it places an emphasis on values, preferences, cultural background, and personal aspirations (Nieuwenhuis et al., 2025).

1.5.1.1 Key Principles

Caring for individuals with MDD difficulties requires collaboration, autonomy, individualized interventions, and adaptability. Care that is both individualized and collaboratively planned and provided by patients, physicians, caregivers, and other support networks (Menear et al., 2022). By allowing individuals to take an active role in making choices about their care, autonomy boosts motivation, self-efficacy, and compliance with long-term care plans (Bravo et al., 2015). It is impossible for generic medication to address people's specific cognitive or physical disabilities, psychosocial circumstances, and functional impairments (Bravo et al., 2015). Care plans for MDD patients need to be flexible enough to accommodate patients' evolving functional abilities as they go through therapy for their depression (Lee et al., 2024).

1.5.1.2 Practical Implementation

Treatment and rehabilitation for MDD should be patient-centered, taking into account the patient's emotional, cognitive, social, and physical functioning as well as their objectives. Culturally responsive interventions take into account the individual's norms, beliefs, and practices. Telemedicine, flexible scheduling, and in-home treatment are just a few of the ways they provide patients more control over their healthcare. These person-centered approaches help people regain agency, speed up the healing process, and enhance their quality of life by going beyond disability treatment that focuses on symptoms or medical conditions.

1.5.2 Trauma-Informed Care

Whether it's the onset, development, or maintenance of MDD, trauma is involved. People with MDD may have worsened emotional dysregulation, cognitive impairment, and social disengagement if they have experienced traumatic experiences, such as interpersonal violence or childhood trauma. Trauma-informed care (TIC) is an essential component of disability-informed treatments that aim to identify trauma, improve safety, and aid in rehabilitation (Radell et al., 2021).

1.5.2.1 Core Principles of Trauma-Informed Care

Trauma-informed MDD disability services put a lot of emphasis on safety, empowerment, cooperation, and being aware of other cultures. All care environments foster physical, emotional, and psychological safety by mitigating threats, judgments, and coercion while ensuring privacy and confidentiality (Mehta et al., 2024). Trauma-informed treatment promotes autonomy, decision-making, and skill development to assist patients in regaining control (Thomas et al., 2019). Care practices prevent re-traumatization by reducing triggers, speaking with care, being open, and being adaptable. Consistent, honest, and caring interactions help people trust each other, and working together to make decisions makes therapy relationships stronger and helps people stick to their treatment plans (Wampold & Flückiger, 2023). Trauma-informed care takes into account a person's culture, history, and gender while making changes to their treatment (Ranjbar et al., 2020).

1.5.2.2 Application to Disability Care in MDD

To include trauma-informed care in MDD disability management, there has to be a full evaluation, therapeutic treatments, changes to the environment, and training for personnel. Trauma history is carefully evaluated, and trauma-informed methodologies are integrated into functional rehabilitation strategies (Cusack et al., 2004). A variety of evidence-based therapies, including mindfulness, eye movement desensitization and reprocessing (EMDR), and trauma-focused cognitive behavioral therapy (TF-CBT), may be used, depending on the patient's needs. As a result of environmental changes, workplaces, social gathering places, and community spaces become more pleasant, predictable, and stress-free (Lawrance et al., 2022). individuals that work with individuals are trauma-informed and educated to be kind, patient, and respectful. By using these strategies in disability care, doctors and support systems may lessen the effects of trauma, get more people to participate in rehabilitation, and speed up functional recovery while still respecting the person's lived experiences.

1.5.3 Strength-Based Support

Disability care has historically focused on deficiencies, limitations, and disabilities. It's important to understand difficulties, but a strength-based approach focuses on recognizing, developing, and employing a person's skills, abilities, and resources to help them get well from MDD. Empowerment, resilience, and self-efficacy are encouraged in recovery-oriented and person-centered methodologies (Ibrahim et al., 2014).

1.5.3.1 Core Elements of Strength-Based Support

Maximizing assets as a means of improving health and functioning, MDD disability treatment identifies and builds upon each patient's unique set of capabilities. Adaptive coping strategies and recalling past triumphs may help people overcome depression. Participation, self-expression, and confidence may all benefit from therapeutic and vocational activities that include creative processes, musical expression, and problem-solving skills. People who have established social networks and abilities may find it easier to reintegrate into social circumstances, experience less isolation, and solicit aid from loved ones. Meaning, motivation, and coping strategies are provided by one's spiritual or cultural foundations, which include one's beliefs, values, and cultural traditions (Padesky & Mooney, 2012).

1.5.3.2 Practical Applications

A comprehensive evaluation of the individual's mental, emotional, social, and practical abilities and limitations is the first step in strengths-based disability therapy for MDD. To aid individuals in accomplishing their objectives, goal planning and intervention design make use of these skills. Some individuals cope with stress by expressing themselves creatively, while others utilize their social abilities to reconnect with community organizations. In order to compensate for functional limitations, skill development builds on these strengths. For example, if one has poor planning or executive functioning, they might compensate by using creative problem-solving. Getting together with others via support groups, cultural or spiritual activities, or volunteering might help you build on your talents. By focusing on what individuals can do instead of what they can't, strength-based strategies boost self-efficacy, motivation, and functional outcomes. To provide full therapy for MDD handicap, they add to medical, trauma-informed, and recovery-focused methods.

1.5.4 Accessibility and Accommodations

MDD often makes it hard to work, go to school, hang out with friends, and live. Disability-informed care focuses on making things easier to get to and making changes so that people can be more independent. According to (Chow et al., 2022), customizing settings, tasks, and supports makes people more involved, productive, and happy.

1.5.4.1 Key Strategies for Accessibility and Accommodations

People who suffer from MDD need to make adjustments to their environment and way of life. By allowing for the adjustment of work or school hours to accommodate fluctuating demands, energy levels, and medical requirements, individuals are able to alleviate stress and remain committed to their treatment. Reducing sensory overload may aid with fatigue, focus issues, and emotional dysregulation by limiting bright lights, loud sounds, and chaos. Regular tasks become simpler with the help of improved executive functioning and less cognitive overload brought about by well-defined schedules and activities. People who have problems focusing, recalling details, or thinking swiftly may benefit from written reminders, visual aids, and explicit instructions. Digital reminders, scheduling software, and gadgets that can be changed all help people become more organized, finish their work, and be more independent. Vocational training, career coaching, and workplace modifications increase the ability of individuals with MDD to maintain or regain employment (Lee et al., 2024).

1.5.4.2 Implementation Considerations

Personalized, collaborative, adaptive, and integrated into larger care concepts are essential for making MDD accessible and adaptable. Individualized care that takes into account a patient's functional profile, therapeutic goals, and personal preferences yields the best results (Tetzlaff et al., 2021). When the individual, their employer, their teachers, and their doctors all work together, they may discover solutions that everyone is happy with. Adaptability also enables modifications to support in reaction to changes in functional capacities or symptoms (Vaillancourt & Newell, 2003). The effectiveness of treatment depends on using a strengths-based, trauma-informed, and person-centered approach (Ranjbar et al., 2020). With these adjustments, persons with MDD have an easier time doing tasks at home and at work, which improves their quality of life, reduces functional impairment, and brings them one step closer to their recovery goals.

1.5.5 Collaborative Multidisciplinary Models

Effective multidisciplinary treatment is available for those with MDD impairments. In order to aid MDD with its physical, mental, social, and emotional issues, a multidisciplinary approach is required. According to (Bertelli et al., 2025), interdisciplinary treatment focuses on healing and is individualized.

1.5.5.1 Key Components of Collaborative Care

A multidisciplinary team is needed for MDD handicap treatment to meet medical, psychological, functional, and social demands. The American Psychiatric Association (2015) states that psychiatrists are able to identify mental health issues, oversee the use of pharmaceuticals, keep track of how symptoms are progressing, and collaborate with other medical professionals to improve the efficacy of both mental health and physical treatment. According to (Curtiss et al., 2021), the field of psychology aims to improve cognitive performance, emotional regulation, and rehabilitation of maladaptive cognition via psychotherapy and psychoeducation. Functional skills are assessed by occupational therapists, who then develop plans to enhance everyday functioning and occupational performance and suggest changes to the environment to encourage participation. Advocates in the field of social work for SDH, community resources, counseling, and case management, as well as for patients' rights. Support workers for persons with disabilities assist those individuals with daily tasks, promote positive behavioral changes, and connect them with educational opportunities, job opportunities, and community events. To improve occupational functioning, vocational counselors help with job planning, skill development, job placement, and workplace adjustments. Peer support experts, on the other hand, encourage recovery by providing social support, mentoring, and practical advice based on personal experience (Quiroz Santos et al., 2025). Patients with MDD may get comprehensive, patient-centered care that is centered on recovery when a multidisciplinary approach is implemented.

1.5.5.2 Benefits of Collaborative Multidisciplinary Care

Treatment for MDD disabilities that involve many disciplines has many advantages. Improving function and reducing fragmented care, combining medical, psychological, and environmental therapy ensures that specialists collaborate. Treatment may be tailored to your requirements and objectives via goal setting and progress tracking. Quality of life, functional improvement, and recovery-oriented treatment are all promoted by the collaborative approach, which also promotes health in all aspects of life (Bond & Drake, 2015).

1.5.5.3 Implementation Considerations

For MDD multidisciplinary therapy to work, everyone has to work together and talk to each other in an orderly way. Regular team meetings or case conferences talk about how the patient is doing, any problems they are having, and any changes that need to be made. Clearly defined roles also help avoid duplication or gaps in treatment. To make sure that care is focused on the person, the person and their family are engaged in making decisions. Evidence-based techniques are included across several disciplines and are customizable to meet particular requirements (Liberman et al., 2001). By combining the knowledge of many experts, collective multidisciplinary approaches provide a comprehensive disability treatment for the wide range of functional impairments experienced by people with MDD (Liberman et al., 2001).

1.6 Key Domains of Disability Care

MDD disability care includes a lot of different things. Table 1.1 shows common problems and good ways to assist individuals keep or get back their functional ability. Individualized, multidisciplinary, person-centered, trauma-informed, and recovery-oriented treatments are most effective (Pinals et al., 2022).

Practical disability-care techniques that are adapted to each category are outlined in Table 1.1, which also identifies important domains in which persons with MDD may have functional impairment.

1.6.1 Activities of Daily Living (ADLs)

To be independent, one must do these basic self-care tasks. People who suffer from major depressive disorder often struggle to carry out even the most basic of everyday tasks because of an overwhelming lack of energy, motivation, or cognitive speed, or as a result of emotionally unstable behavior. Functional impairment and worsening depression may result from problems in this area (Troyer, 2018).

Table 1.1 Examples of disability-related challenges in MDD and corresponding care strategies

S. No	Domain of functioning	Common challenges in MDD	Examples of disability care strategies
1	Activities of daily living (ADLs)	Neglect of hygiene; irregular meals; disrupted sleep	Structured routines, prompts, environmental cues, graded task practice
2	Instrumental ADLs (IADLs)	Difficulty managing finances, appointments, medication	Checklists, reminders, budgeting assistance, medication support
3	Cognition	Poor concentration, slowed thinking, impaired planning	Cognitive aids, task simplification, reduced sensory load, supportive supervision
4	Social participation	Isolation, reduced communication, avoidance	Peer support groups, social skills training, gradual exposure
5	Work/Study	Absenteeism, presenteeism, reduced productivity	Flexible hours, job coaching, workload modification
6	Emotional regulation	Persistent sadness, irritability, hopelessness	Psychotherapy, mood tracking, coping skills training
7	Physical/ Somatic	Fatigue, pain, psychomotor slowing	Energy conservation strategies, graded exercise, pacing

1.6.1.1 Common Challenges in ADLs

MDD functional deficiencies frequently have an effect on daily life skills. Not bathing, grooming, or taking care of your teeth are all examples of poor hygiene. Problems with nutrition and meal planning might include inconsistent meals, a lack of appetite, and not getting enough nutrients. Sleep hygiene is commonly broken, which may lead to insomnia or hypersomnia. MDD also makes everyday tasks like cleaning, doing laundry, and other domestic chores harder.

1.6.1.2 Support Strategies

To help with ADLs, effective disability-informed MDD treatments build skills, set up routines, and modify the environment. Scaffolding gives people prompts, checklists, or guided help to help them gain skills in personal care, food preparation, and managing their homes. Organizing living spaces, using visual cues, and making work sequences easier to follow all help reduce complexity and mental strain. Occupational therapy, cognitive therapy, and programs that provide psychosocial support all include these tactics. Restoring autonomy and self-efficacy and reducing the negative impact of functional impairment on mood and general well-being may be achieved by providing structured help for ADL inadequacies to individuals with MDD.

1.6.2 *Instrumental Activities of Daily Living (IADLs)*

IADLs are higher-level tasks that are necessary for independence and engagement in the community. People with MDD find it simpler to stop doing these activities because they have trouble thinking, are not motivated, have trouble with executive function, and are tired. IADL deficiencies restrict autonomy and augment reliance on support systems (Fish, 2018).

1.6.2.1 Common Challenges in IADLs

MDD may make IADLs, which are very important for independence, quite bad. It could be hard to keep track of your spending, pay your bills, and make a budget. It may also be hard to establish plans for appointments, meetings, and everyday tasks when calendars are messy. Not taking the right amount of medicine, not taking it at the right times, or not following the instructions on the prescription might all make it harder to stick to a drug regimen. It could be hard to find a way to go to work, an appointment, or run an errand. MDD may impede cleaning, organizing, and upkeep of the home (Benge et al., 2024).

1.6.2.2 Support Strategies

Assistance from others, consistent monitoring, and continuous skill growth are necessary for managing MDD IADL deficiency. Habits are more easily maintained with the use of calendars, alarms, applications for smartphones, and notes. Step-by-step planning and checklists make hard jobs easier and lighten the load on your brain. As skills and confidence grow, graduated task assistance gives less help. Cognitive and occupational therapy are used with IADL training, executive function exercises, memory aids, and adjustments to the environment (Dawson et al., 2015). By supporting people with MDD with their IADLs, caregivers may help them become more independent, improve their quality of life, and reduce their functional impairment.

1.6.3 *Psychosocial Rehabilitation (PSR)*

PSR is part of MDD disability therapy. It helps people with MDD develop their social skills, take part in their community, and feel better about themselves. Unlike treatments that concentrate on symptoms (Farkas, 2013), PSR deals with functional impairments that make it hard to do meaningful things and interact with other people.

1.6.3.1 Key Goals of Psychosocial Rehabilitation

MDD may make you less confident, less able to talk to others, and less able to do things that are important in social and communal situations. Interventions help people feel better about themselves, their social skills, and their ability to get along with others. To make it easier to become involved in social activities, verbal and nonverbal communication skills are developed. Having a job, going to school, volunteering, or being part of a community group may give you a feeling of connection and purpose. To increase quality of life and well-being, people are also urged to do things like hobbies, leisure, and creative pursuits.

1.6.3.2 Programmatic Approaches

In MDD PSR programs, structured group activities, mentorship, training in social skills, and inclusion into the community are all frequent. Recreation, workshops, and peer support groups help people make friends, work together, and feel like they belong. Peer support and mentorship provide you help, role models, and ways to deal with problems from people who know what they're doing. Relationships, assertiveness, communication, and conflict resolution are all enhanced by social skills training. People with social isolation or anxiety may find relief and more meaningful connections via community integration programs, which help them adjust to public and social settings over time with organized assistance (Parikh et al., 2016).

1.6.3.3 Practical Considerations

Individual differences in culture, functional requirements, and interests should inform the development of a PSR plan for MDD patients. Peer specialists, psychologists, and occupational therapists all work to improve functional capacity. PSR is most effective when integrated with medical and psychiatric therapy within a holistic, recovery-oriented framework. These tailored therapies address social and functional deficiencies to facilitate recovery, engagement, and reduce isolation in people with MDD (Dixon et al., 2001).

1.6.4 Vocational Support

Personal growth, autonomy, and emotional well-being are all positively impacted by having a job. Absenteeism, presenteeism, cognitive impairments, and diminished enthusiasm to work are all symptoms of major depressive disorder (MDD). Hence, vocational assistance is a need of disability care in order to help people with disabilities retain or recover meaningful work and actively participate in it (Hoffmann et al., 2014).

1.6.4.1 Key Vocational Challenges in MDD

MDD may hinder attendance and productivity at work. Issues with focus, making decisions, and executive functioning make it hard to get a job. Fear of losing your job and stigma may also make it hard for people to work together and participate. MDD may also make it harder to get a job, keep it, and move forward in your career (Williams et al., 2016).

1.6.4.2 Support Strategies

Practical skills, adjustments in the environment, and psychological support all help MDD vocational rehabilitation work better. Personal job coaching helps people deal with stress, bad habits, and what is expected of them at work. Employment programs help people get employment and provide them help while they work. Getting assistance with your résumé, interview, and career planning will help you get a job and show off your skills. To alleviate stress and maximize productivity, it is best to work in an atmosphere that is both peaceful and has a lot of leeway in terms of scheduling and workload. To ensure long-term employment, it is recommended to gradually return to work in order to balance energy, mental health, and workload.

1.6.4.3 Implementation Considerations

A person's strengths, areas for improvement, and professional goals should all be considered during vocational rehabilitation for MDD. Assistance that is practical, structured, and applicable to the workplace is given to individuals by mental health professionals, vocational counselors, and employers in a collaborative effort that helps them succeed. Vocational aid, cognitive remediation, psychosocial rehabilitation, and strength-based techniques all help people have better functional outcomes. Structured and focused vocational assistance may improve the occupational functioning, self-efficacy, and long-term rehabilitation of individuals with Major Depressive Disorder (Bond et al., 2001).

1.6.5 Crisis Management

Crisis management is an element of treating MDD impairment, especially for those who are at risk of suicide, a big drop in function, or sudden worsening of symptoms. Planning well makes ensuring that people are safe, that they get the care they need, and that they don't get hurt at times when they are most at risk.

1.6.5.1 Key Areas Requiring Crisis Planning

It is crucial to identify and treat suicidal thoughts or self-harm as soon as possible since they might trigger MDD crises. By making it difficult to perform both essential and instrumental ADLs, functional breakdown due to a crisis might put safety and autonomy at risk. Not taking psychiatric drugs as prescribed, skipping doses, and having bad side effects might make symptoms and functioning worse. Breakdowns in relationships, money problems, and stressful life events may make depression symptoms worse and make it harder to cope. This means that treatment has to be rapid and organized to get people back to normal and stop the problems from becoming worse.

1.6.5.2 Crisis Management Strategies

During high-risk moments, safety and well-organized help are very important for managing MDD crises. Make personal safety plans that include triggers, ways to deal with them, and useful contacts to lower the risk. Individual-caregiver safety contracts may lead to crisis measures. Mental health practitioners, crisis hotlines, and local emergency services are all important for a quick reaction. Advanced directives provide treatment options to make sure that people may keep their independence and continuity in really bad situations. Caregivers and experts may safely intervene in circumstances thanks to fast-response solutions.

1.6.5.3 Implementation Considerations

In MDD, crisis management comprises collaboration, evaluation, and adjustment. Disability benefits, psychological rehabilitation, and therapeutic treatments for the long term may all benefit from a combination of crisis interventions and continuous care. Improving the effectiveness of safety and intervention measures may be achieved via training loved ones, caregivers, and support personnel to identify and handle emergencies. Adding crisis management to disability care helps doctors and support team's lower risks, keep things stable, keep patients safe, and assist patients with MDD get better, become more independent, and feel better overall (Karam et al., 2021).

1.7 The Role of Family and Caregivers

Individuals with MDD, especially those with disabilities, require familial and caregiving support. Participation enhances rehabilitation, quality of life, and adherence to therapy. Caring for someone involves overcoming financial, social, and emotional obstacles (Kopelowicz & Liberman, 2003).

1.7.1 Understanding the Caregiver Experience

Taking care of someone who suffers from MDD is no easy feat. When symptoms change, caregivers experience feelings of anxiety, fear, and helplessness. Uncertainty about symptom trajectories, treatment efficacy, and functional recovery adds stress to caring. Long-term responsibilities such as treatment costs, diminished work capacity, and caregiving may lead to compassion fatigue and financial strain. Because caregivers who are too busy can't always help, it's important to recognize and deal with caregiver strain in order to provide long-term and effective disability care (Cheng et al., 2022).

1.7.2 Education and Psychoeducation

By disseminating accurate information, caregivers may aid patients in their recovery and maintenance of good health. It is critical for caregivers to recognize the warning signs of depression, thoughts of suicide, and loss of function, and to be able to discuss these issues without passing judgment. People may better prepare for danger and get expert assistance with crisis response training. To avoid burnout, it's important to learn self-care skills like setting limits and making time for relaxation and healthy living. Psychoeducation may help caregivers deal with obstacles, support treatment objectives, and reduce stress for themselves and their MDD patients (Gutiérrez-Maldonado & Caqueo-Urízar, 2007).

1.7.3 Collaborative Support

Management of Major Depressive Disorder (MDD) therapy and functional recovery are improved when family engagement is balanced with personal autonomy. Patients and caregivers work together effectively when they are involved in treatment decision-making, goal-setting, and problem-solving. By coordinating treatment with practical goals, professions may inspire people to eat well, get enough sleep, exercise, and spend time with friends and family. Evidence suggests that families of people with severe depressive disorder may have an important role in rehabilitation, functional results, and social and emotional resilience (Dixon et al., 2001).

1.8 Stigma, Inclusion, and Human Rights

Symptom relief and functional improvement are only the beginning of what a patient with MDD may expect from treatment. Furthermore, it works to remove social barriers, encourage inclusivity, and protect human rights. It is more challenging to recover, participate, and function when faced with stigma and prejudice. (Marx et al., 2023) assert that rights-based and inclusive approaches are essential components of comprehensive treatment.

1.8.1 Stigma as a Barrier

The internal and external stigmatization of mental illness makes it even more difficult for people with MDD to engage in daily activities and form meaningful relationships. Seeking mental health treatment might be intimidating for some individuals. Stigmatization in the workplace, schools, and society at large impedes independence. Feelings of shame may make it harder to overcome poor self-esteem, despair, and begin the rehabilitation process. Reducing stigma improves treatment, functioning, and social participation for people with MDD (Link et al., 2001).

1.8.2 Anti-Stigma Strategies

More education, individual involvement, and systemic advocacy are necessary steps in eliminating the stigma associated with MDD. Common misunderstandings, how it works, and therapy for MDD are all part of the educational process. Meeting with MDD survivors reduces stigma and social isolation. Media balance and empathy fight harmful misconceptions and help people understand each other. Lobbying for community, organizational, and policy changes makes things more inclusive. These methods could help care systems and communities decrease barriers, get more people involved, and enhance the psychosocial outcomes of MDD patients.

1.8.3 Rights-Based Approaches

The CRPD, or UN Convention on the Rights of Persons with Disabilities, helps people with MDD. People with disabilities have the right to an inclusive community that welcomes them and supports their cultural, civic, and recreational pursuits; the right to an education; the right to an equal opportunity job; the right to receive health care, social services, and reasonable accommodations in the form of modifications to the workplace or school; and the right to an environment free from discrimination. Autonomy in treatment, openness about treatment, mental health treatments, psychosocial rehabilitation, and care that takes disabilities into account are all guarantees of informed consent. Disability care that is based on MDD rights is ethical, equitable, and conforms to global norms for respect for human dignity and inclusion.

1.9 Assessment and Care Planning

It is crucial to plan and assess disability care for MDD. Methods like this allow for continuous monitoring and modification of therapies based on functional needs, strengths, and rehabilitation objectives (Lee et al., 2024).

1.9.1 Comprehensive Functional Assessment

Comprehensive functional assessments analyze numerous areas to adjust MDD treatment. Significantly important are cognitive functioning, occupational capacity, and the ability to live independently. Cognitive functioning includes things like attention, memory, executive functions, and processing speed. Occupational capacity includes things like job performance, vocational talents, and workplace engagement. Psychosocial functioning, life satisfaction, subjective well-being, support networks, community involvement, and interpersonal relationships are some of the factors examined in the research. Underlying disability-informed treatment is objective, patient-centered data obtained from organized interviews, direct observation, standardized scales, and self-report evaluations.

1.9.2 Goal-Oriented Planning

The results of an MDD examination should be transformed into attainable objectives via systematic, recovery-oriented care planning. Setting realistic short-term goals will help you feel more confident. Long-term rehabilitation plans should be based on your values, functional goals, and quality of life. To measure progress and interventions, you need to have quantifiable results. When modifying care plans, think about symptoms, how well the individual can function, and what is most important to them. Disability care is useful, relevant, and tailored to the person's strengths and weaknesses via goal-oriented planning.

1.9.3 Tracking Progress

Patients with MDD are followed to see how well the therapy is working and to help plan the next steps. Mood charts and activity diaries are useful for keeping track of changes in mood and symptoms of depression every day or week. The results of an MDD examination should be transformed into attainable objectives via systematic, recovery-oriented care planning. By using both objective indicators and patient-reported outcomes, care teams may improve treatments, engagement, and functional improvements that are focused on recovery.

1.10 Evidence-Based Treatment Within Disability Care

Treatment for MDD disabilities must include both practical assistance and evidence-based treatment for symptoms, as well as cognitive and psychosocial functioning. Treatments that are disability-informed are those that are realistic, person-centered, and rehabilitation-oriented (Kopelowicz & Liberman, 2003).

1.10.1 Psychotherapy

Psychotherapy is necessary for disability rehabilitation and MDD treatment. CBT deals with negative thoughts and actions, problem-solving, and functional gains that are goal-oriented. In order to combat withdrawal and improve mood and motivation, behavioral activation promotes engaging in rewarding activities. A person's social functioning, role transitions, and issues relating to other people are the focus of Interpersonal Therapy (IPT). While psychodynamic therapy looks at patterns and dynamics in relationships that can be limiting functioning, acceptance, and mindfulness practices like ACT and mindfulness improve psychological flexibility, emotional regulation, and stress management. For psychotherapy to be effective, it must address the cognitive, motivational, and functional abilities of patients with Major Depressive Disorder (Parikh et al., 2016).

1.10.2 Medication Management

Drugs and people who work with disabled people may help with depression. Reminders, monitoring routines, and adherence methods help people stick to their medicine. Side effects are watched and reported to the prescriber. Patients, psychiatrists, and other care providers are taught about how medications work, their advantages, and their dangers. Coordination is also easier.

1.10.3 Somatic Therapies

The somatic therapies available for severe, chronic, or treatment-resistant depression include ECT, a non-invasive neuromodulation technique, TMS, a fast-acting antidepressant, and ketamine or esketamine therapy. ECT is particularly useful for cases of extreme depression or suicidality. (Łysik et al., 2025) states that disability care may help with transportation, monitoring side effects, planning aftercare, and returning to regular activities after treatment.

1.10.4 Holistic and Lifestyle Supports

Lifestyle treatment could help people with MDD become more resilient, operate better, and have a better quality of life. Mood, energy, and cognitive function are all improved with regular exercise; circadian rhythms are stabilized with good sleep hygiene; physical and mental health are enhanced with a balanced diet; stress is reduced with relaxation, mindfulness, and coping mechanisms; and emotional expression and social engagement are fostered through creative arts therapies. According to (Ramar et al., 2021), care teams can improve patients' recovery and ability to participate in daily activities by implementing lifestyle interventions that take disability into account, as well as by using evidence-based psychotherapy, pharmaceutical treatments, and somatic therapies.

1.11 Ethical Considerations in Disability Care

In MDD disability care, it is important to respect people's rights, dignity, and individuality. (Doody et al., 2023) say that person-centered, culturally sensitive, and lawful methods protect both patients and caregivers.

1.11.1 Autonomy and Consent

In disability care ethics, it is important to respect people's freedom and provide people with MDD the power to choose their own treatment and assistance. This includes taking the time to explain all of the potential side effects, advantages, and risks of a treatment choice in a way that anybody can comprehend; including the patient in decision-making processes; and promoting self-determination by honoring the patient's values and wishes. Potentially increasing participation, intervention adherence, and functional recovery is the goal of care professionals who provide patients more autonomy.

1.11.2 Confidentiality

For trust and a healthy therapeutic collaboration, family or supported living has to be kept secret. You should only give out information on symptoms, treatment, and personal details if you have permission or if it is legally required for safety. To protect privacy and coordinate interventions, you need to keep things secret, safe, and work together.

1.11.3 Cultural Safety

Culture, identity, and community all have a role in a person's mental health, therefore it's crucial for disability care providers to be culturally competent and sensitive. Acknowledging and honoring the client's cultural beliefs and values; communicating with the client in their preferred language or through interpreters; understanding community norms, such as group decision-making, family duties, and social expectations; and respecting the client's ethnic, gender, sexual, and spiritual identities. By encouraging confidence, involvement, and substantial involvement, culturally safe care improves health outcomes and patients' quality of life (Curtis et al., 2019).

1.11.4 Professional Boundaries

Everyone from clients to caregivers to therapeutic couples may benefit from clear boundaries in the workplace. Among these tasks is the establishment of protocols for dealing with boundary issues that are either complicated or pose a high risk, the definition of roles and obligations, and the avoidance of any potential compromise to safety or objectivity caused by parallel connections. Professional, safe, and considerate of individuals' autonomy, MDD disability care is the best option.

1.12 Technology and Innovation in Disability Care

Advances in technology have made MDD disability care more accessible, personalized, and autonomous. (Torous et al., 2025) found that digital technology and assistive gadgets may be used in care frameworks to help with functional rehabilitation, monitoring, and engagement.

1.12.1 Digital Therapeutics

Enhancing mental health and functioning may be achieved via digital therapies, which include smartphone apps, web platforms, and online programs. Depression, sleep, and activity tracking; cognitive behavioral therapy (CBT) exercises for rewiring thoughts and actions; medication, appointment, and daily routine reminders; and relaxation or mindfulness programs for controlling emotions and stress are common elements. With these aids, individuals with disabilities may take an active role in their own care while still receiving disability-sensitive medical attention.

1.12.2 Assistive Technologies

Individuals with MDD may find that technological aids allow them to perform more independently and improve their functional capacity. Automatic lighting, reminders, and environmental controls are smart home features that assist with ADLs. Helpful for IADLs and employment are scheduling and organizing applications such as to-do lists, alerts, and planning tools. Sleep, exercise, stress, and heart rate may all be tracked by wearable health gadgets. When these technologies eliminate common issues, people feel more empowered, confident, and have a higher quality of life (Harvey et al., 2022).

1.12.3 Telehealth

Patients with impairments, especially those in rural locations, who have mobility issues, or who suffer from social anxiety, may now get disability-specific therapy via telehealth. Many people use it for things like online peer support groups to combat isolation and boost social engagement, virtual case management to organize treatment and monitor functional objectives, and remote psychotherapy. Improved continuity of treatment, access to specialists, and interdisciplinary support for in-person therapy are all benefits of telehealth. According to (Nowrouzi-Kia et al., 2025), people with Major Depressive Disorder may benefit from telehealth, digital treatments, and assistive technology in terms of participation, autonomy, and recovery.

1.13 Future Directions

Rapid advancements in neuroscience, psychological studies, and public understanding of mental health issues are altering the therapeutic landscape for MDD. According to recent trends (Marx et al., 2023), functional assistance, recuperation, and inclusion are all affected.

1.13.1 Key Emerging Trends

Personalized, integrated, and socially responsive disability-informed MDD therapy is on the rise. Genetic and neuroimaging biomarkers in personalized psychiatry adapt treatments, forecast treatment efficacy, and improve functionality. Collaborative care approaches bring together mental health and general care to help people stay well and operate well. Peer workers with lived experience care for, mentor, and advocate for others, which increases involvement and social support. Functional impairment and rehabilitation treatments deal with things like poverty, unstable housing, being alone, and discrimination. Legal and regulatory changes make it easier for people with disabilities to get jobs and fight discrimination. According to (Stein et al., 2022), wellness programs, flexible scheduling, task modifications, regulations regarding mental health leave, and other workplace adaptations encourage job involvement and decrease presenteeism. This, in turn, aids in long-term rehabilitation and functional reintegration.

1.13.2 Implications for Practice

Future disability-informed MDD therapy includes integration, individualization, and recovery. More personalized treatment programs will include biological, psychological, social, and technological inputs. To meet functional goals and recovery goals, therapy will focus on rights-based, autonomy-based, and empowerment-based methods. Collaborative problem-solving is essential for generating novel ideas in the fields of clinical psychology, social work, vocational counseling, policy advocacy, and peer specialists. People with major depressive disorder may see improvements in their functional outcomes, quality of life, and social engagement as a result of more precise, thorough, and all-encompassing disability treatment (Cook et al., 2025).

1.14 Conclusions

Functional impairment is a consequence of the emotional, cognitive, social, occupational, and physical aspects impacted by MDD. Because of this diversity of issues, therapy for MDD impairment must be evidence-based and recovery-focused. People are able to gain more autonomy, self-determination, and life satisfaction via the integration of bio-psycho-social factors, which include medical, psychological, and social aspects. The most effective treatments include multidisciplinary teams that prioritize the needs of each individual patient while also taking trauma into account, ensuring cultural safety, and encouraging active participation from patients' loved ones and friends. Gains in functionality, participation, and social inclusion may be achieved via the elimination of stigma, the defense of human rights, and the implementation of reasonable accommodations. In order to help people with MDD recover and participate in daily life, interdisciplinary approaches and individualized treatment plans boost autonomy, respect, and health.

References

Alegría, M., Alvarez, K., Cheng, M., & Falgas-Bague, I. (2023). Recent advances on social determinants of mental health: Looking fast forward. *American Journal of Psychiatry, 180*(7), 473–482.

Beck, A. T., & Haigh, E. A. (2014). Advances in cognitive theory and therapy: The generic cognitive model. *Annual Review of Clinical Psychology, 10*(1), 1–24.

Benge, J. F., Ali, A., Chandna, N., Rana, N., Mis, R., González, D. A., Hilsabeck, R. C., et al. (2024). Technology-based instrumental activities of daily living in persons with Alzheimer's disease and related disorders. *Alzheimer's & Dementia: Diagnosis, Assessment & Disease Monitoring, 16*(4), Article e70022.

Berk, M., Köhler-Forsberg, O., Turner, M., Penninx, B. W., Wrobel, A., Firth, J., Marx, W., et al. (2023). Comorbidity between major depressive disorder and physical diseases: A comprehensive review of epidemiology, mechanisms and management. *World Psychiatry, 22*(3), 366–387.

Bertelli, M. O., Bianco, A., Salerno, L., & Salvador-Carulla, L. (2025). Integrated Care for People with Intellectual Disability and Autism Spectrum Disorder. In *Handbook of Integrated Care* (pp. 1037–1056). Springer Nature Switzerland.

Boes, A. D., Kelly, M. S., Trapp, N. T., Stern, A. P., Press, D. Z., & Pascual-Leone, A. (2018). Noninvasive brain stimulation: Challenges and opportunities for a new clinical specialty. *The Journal of Neuropsychiatry and Clinical Neurosciences, 30*(3), 173–179.

Bond, G. R., & Drake, R. E. (2015). The critical ingredients of assertive community treatment. *World Psychiatry, 14*(2), 240.

Bond, G. R., Becker, D. R., Drake, R. E., Rapp, C. A., Meisler, N., Lehman, A. F., Blyler, C. R., et al. (2001). Implementing supported employmentas an evidence-based practice. *Psychiatric Services, 52*(3), 313–322.

Bottema-Beutel, K., Kapp, S. K., Lester, J. N., Sasson, N. J., & Hand, B. N. (2020). Avoiding ableist language: Suggestions for autism researchers. *Autism in Adulthood, 3*(1), 18–29.

Bravo, P., Edwards, A., Barr, P. J., Scholl, I., Elwyn, G., & McAllister, M. (2015). Conceptualising patient empowerment: A mixed methods study. *BMC Health Services Research, 15*(1), 1–14.

Carey, E., Ryan, R., Sheikhi, A., & Dore, L. (2023). Exercising autonomy—The effectiveness and meaningfulness of autonomy support interventions engaged by adults with intellectual disability. A mixed-methods review. *British Journal of Learning Disabilities, 51*(3), 307–323.

Chakrabarty, T., Hadjipavlou, G., & Lam, R. W. (2016). Cognitive dysfunction in major depressive disorder: Assessment, impact, and management. *Focus, 14*(2), 194–206.

Cheng, W. L., Chang, C. C., Griffiths, M. D., Yen, C. F., Liu, J. H., Su, J. A., Pakpour, A. H., et al. (2022). Quality of life and care burden among family caregivers of people with severe mental illness: Mediating effects of self-esteem and psychological distress. *BMC Psychiatry, 22*(1), 672.

Chow, T. K., Bowie, C. R., Morton, M., Lalovic, A., McInerney, S. J., & Rizvi, S. J. (2022). Contributors of functional impairment in major depressive disorder: A biopsychosocial approach. *Current Behavioral Neuroscience Reports, 9*(2), 59–72.

Cook, A., Hunt, R., Silcox, J., Canas, E., & MacDougall, A. G. (2025). Defining youth-centred practice in mental health care. *BMC Psychiatry, 25*(1), 578.

Cuijpers, P., Miguel, C., Harrer, M., Plessen, C. Y., Ciharova, M., Ebert, D., & Karyotaki, E. (2023). Cognitive behavior therapy vs. control conditions, other psychotherapies, pharmacotherapies and combined treatment for depression: A comprehensive meta-analysis including 409 trials with 52,702 patients. *World Psychiatry, 22*(1), 105–115.

Curtis, E., Jones, R., Tipene-Leach, D., Walker, C., Loring, B., Paine, S. J., & Reid, P. (2019). Why cultural safety rather than cultural competency is required to achieve health equity: A literature review and recommended definition. *International Journal for Equity in Health, 18*(1), 174.

Curtiss, J. E., Levine, D. S., Ander, I., & Baker, A. W. (2021). Cognitive-behavioral treatments for anxiety and stress-related disorders. *Focus, 19*(2), 184–189.

Cusack, K. J., Frueh, B. C., & Brady, K. T. (2004). Trauma history screening in a community mental health center. *Psychiatric Services, 55*(2), 157–162.

Dawson, A., Bowes, A., Kelly, F., Velzke, K., & Ward, R. (2015). Evidence of what works to support and sustain care at home for people with dementia: A literature review with a systematic approach. *BMC Geriatrics, 15*(1), 59.

Dixon, L., McFarlane, W. R., Lefley, H., Lucksted, A., Cohen, M., Falloon, I., Sondheimer, D., et al. (2001). Evidence-based practices for services to families of people with psychiatric disabilities. *Psychiatric Services, 52*(7), 903–910.

Doody, O., Hennessy, T., Moloney, M., Lyons, R., & Bright, A. M. (2023). The value and contribution of intellectual disability nurses/nurses caring for people with intellectual disability in intellectual disability settings: A scoping review. *Journal of Clinical Nursing, 32*(9–10), 1993–2040.

Ee, C., Lake, J., Firth, J., Hargraves, F., De Manincor, M., Meade, T., Sarris, J., et al. (2020). An integrative collaborative care model for people with mental illness and physical comorbidities. *International Journal of Mental Health Systems, 14*(1), 83.

Evans-Lacko, S., & Knapp, M. (2016). Global patterns of workplace productivity for people with depression: Absenteeism and presenteeism costs across eight diverse countries. *Social Psychiatry and Psychiatric Epidemiology, 51*(11), 1525–1537.

Farkas, M. (2013). Introduction to psychiatric/psychosocial rehabilitation (PSR): History and foundations. *Current Psychiatry Reviews, 9*(3), 177–187.

Fischhoff, B., & Broomell, S. B. (2020). Judgment and decision making. *Annual Review of Psychology, 71*(1), 331–355.

Fish, J. (2018). Lawton-Brody instrumental activities of daily living scale. In *Encyclopedia of clinical neuropsychology* (pp. 1966–1967). Springer.

Fredrick, J. W., & Becker, S. P. (2023). Cognitive disengagement syndrome (sluggish cognitive tempo) and social withdrawal: Advancing a conceptual model to guide future research. *Journal of Attention Disorders, 27*(1), 38–45.

Gaynes, B. N., Lux, L., Gartlehner, G., Asher, G., Forman-Hoffman, V., Green, J., Lohr, K. N., et al. (2020). Defining treatment-resistant depression. *Depression and Anxiety, 37*(2), 134–145.

Ghanean, H., Ceniti, A. K., & Kennedy, S. H. (2018). Fatigue in patients with major depressive disorder: Prevalence, burden and pharmacological approaches to management. *CNS Drugs, 32*(1), 65–74.

Gonda, X., Pompili, M., Serafini, G., Carvalho, A. F., Rihmer, Z., & Dome, P. (2015). The role of cognitive dysfunction in the symptoms and remission from depression. *Annals of General Psychiatry, 14*(1), 27.

Gustavson, K., Knudsen, A. K., Nesvåg, R., Knudsen, G. P., Vollset, S. E., & Reichborn-Kjennerud, T. (2018). Prevalence and stability of mental disorders among young adults: Findings from a longitudinal study. *BMC Psychiatry, 18*(1), 65.

Gutiérrez-Maldonado, J., & Caqueo-Urízar, A. (2007). Effectiveness of a psycho-educational intervention for reducing burden in Latin American families of patients with schizophrenia. *Quality of Life Research, 16*(5), 739–747.

Hames, J. L., Hagan, C. R., & Joiner, T. E. (2013). Interpersonal processes in depression. *Annual Review of Clinical Psychology, 9*(1), 355–377.

Hamilton, J. P., Etkin, A., Furman, D. J., Lemus, M. G., Johnson, R. F., & Gotlib, I. H. (2012). Functional neuroimaging of major depressive disorder: A meta-analysis and new integration of baseline activation and neural response data. *American Journal of Psychiatry, 169*(7), 693–703.

Harvey, A. G. (2011). Sleep and circadian functioning: Critical mechanisms in the mood disorders? *Annual Review of Clinical Psychology, 7*(1), 297–319.

Harvey, P. D., Depp, C. A., Rizzo, A. A., Strauss, G. P., Spelber, D., Carpenter, L. L., Torous, J., et al. (2022). Technology and mental health: State of the art for assessment and treatment. *American Journal of Psychiatry, 179*(12), 897–914.

Hinshaw, S. P., & Stier, A. (2008). Stigma as related to mental disorders. *Annual Review of Clinical Psychology, 4*(1), 367–393.

Hoffmann, H., Jäckel, D., Glauser, S., Mueser, K. T., & Kupper, Z. (2014). Long-term effectiveness of supported employment: 5-year follow-up of a randomized controlled trial. *American Journal of Psychiatry, 171*(11), 1183–1190.

Ibrahim, N., Michail, M., & Callaghan, P. (2014). The strengths based approach as a service delivery model for severe mental illness: A meta-analysis of clinical trials. *BMC Psychiatry, 14*(1), 243.

IsHak, W. W., James, D. M., Mirocha, J., Youssef, H., Tobia, G., Pi, S., Cohen, R. M., et al. (2016). Patient-reported functioning in major depressive disorder. *Therapeutic Advances in Chronic Disease, 7*(3), 160–169.

Jacobson, N., & Greenley, D. (2001). What is recovery? A conceptual model and explication. *Psychiatric Services, 52*(4), 482–485.

Karam, M., Chouinard, M. C., Poitras, M. E., Couturier, Y., Vedel, I., Grgurevic, N., & Hudon, C. (2021). Nursing care coordination for patients with complex needs in primary healthcare: A scoping review. *International Journal of Integrated Care, 21*(1), 16.

Kennedy, S. H., Lam, R. W., McIntyre, R. S., Tourjman, S. V., Bhat, V., Blier, P., & CANMAT Depression Work Group et al. (2016). Canadian Network for Mood and Anxiety Treatments (CANMAT) 2016 clinical guidelines for the management of adults with major depressive disorder: section 3. Pharmacological treatments. *The Canadian Journal of Psychiatry, 61*(9), 540–560.

Kessler, R. C., & Bromet, E. J. (2013). The epidemiology of depression across cultures. *Annual Review of Public Health, 34*(1), 119–138.

Kim, S., Jang, Y. S., & Park, E. C. (2025). Associations between social isolation, withdrawal, and depressive symptoms in young adults: A cross-sectional study. *BMC Psychiatry, 25*(1), 1–12.

King, L. A., & Hicks, J. A. (2021). The science of meaning in life. *Annual Review of Psychology, 72*(1), 561–584.

Kirkbride, J. B., Anglin, D. M., Colman, I., Dykxhoorn, J., Jones, P. B., Patalay, P., Griffiths, S. L., et al. (2024). The social determinants of mental health and disorder: Evidence, prevention and recommendations. *World Psychiatry, 23*(1), 58–90.

Kopelowicz, A., & Liberman, R. P. (2003). Integration of care: Integrating treatment with rehabilitation for persons with major mental illnesses. *Psychiatric Services, 54*(11), 1491–1498.

Lagerveld, S. E., Bültmann, U., Franche, R. L., Van Dijk, F. J. H., Vlasveld, M. C., van der Feltz-Cornelis, C. M., Nieuwenhuijsen, K., et al. (2010). Factors associated with work participation and work functioning in depressed workers: A systematic review. *Journal of Occupational Rehabilitation, 20*(3), 275–292.

Lawrance, E. L., Thompson, R., Newberry Le Vay, J., Page, L., & Jennings, N. (2022). The impact of climate change on mental health and emotional wellbeing: A narrative review of current evidence, and its implications. *International Review of Psychiatry, 34*(5), 443–498.

Lee, M., Choi, H., & Jo, Y. T. (2024). Targeting emotion dysregulation in depression: An intervention mapping protocol augmented by participatory action research. *BMC Psychiatry, 24*(1), 595.

Leiter, M. P., & Maslach, C. (1988). The impact of interpersonal environment on burnout and organizational commitment. *Journal of Organizational Behavior, 9*(4), 297–308.

Liberman, R. P., Hilty, D. M., Drake, R. E., & Tsang, H. W. (2001). Requirements for multidisciplinary teamwork in psychiatric rehabilitation. *Psychiatric Services, 52*(10), 1331–1342.

Lieberz, J., Shamay-Tsoory, S. G., Saporta, N., Esser, T., Kuskova, E., Stoffel-Wagner, B., Scheele, D., et al. (2021). Loneliness and the social brain: How perceived social isolation impairs human interactions. *Advanced Science, 8*(21), 2102076.

Link, B. G., Struening, E. L., Neese-Todd, S., Asmussen, S., & Phelan, J. C. (2001). Stigma as a barrier to recovery: The consequences of stigma for the self-esteem of people with mental illnesses. *Psychiatric Services, 52*(12), 1621–1626.

Łysik, A., Logoń, K., Szczygieł, A., Wołoszczak, J., Wrześniewska, M., & Leszek, J. (2025). Innovative approaches in the treatment-resistant depression: exploring different therapeutic pathways. *GeroScience*, 1–16.

Marx, W., Manger, S. H., Blencowe, M., Murray, G., Ho, F. Y. Y., Lawn, S., O'Neil, A., et al. (2023). Clinical guidelines for the use of lifestyle-based mental health care in major depressive disorder: World Federation of Societies for Biological Psychiatry (WFSBP) and Australasian Society of Lifestyle Medicine (ASLM) taskforce. *The World Journal of Biological Psychiatry, 24*(5), 333–386.

McDowell, C., & Fossey, E. (2015). Workplace accommodations for people with mental illness: A scoping review. *Journal of Occupational Rehabilitation, 25*(1), 197–206.

Mehra, A., Khanna, J., Singh, G., Sachdeva, V., & Bedi, N. (2025). A Comprehensive Review on Major Depressive Disorder: Exploring Etiology, Pathogenesis and Clinical Approaches. *Current Behavioral Neuroscience Reports, 12*(1), 18.

Mehta, L. S., Churchwell, K., Coleman, D., Davidson, J., Furie, K., Ijioma, N. N., Shanafelt, T., et al. (2024). Fostering psychological safety and supporting mental health among cardiovascular health care workers: A science advisory from the american heart association. *Circulation, 150*(2), e51–e61.

Melillo, A., Sansone, N., Allan, J., Gill, N., Herrman, H., Cano, G. M., Galderisi, S., et al. (2025). Recovery-oriented and trauma-informed care for people with mental disorders to promote human rights and quality of mental health care: A scoping review. *BMC Psychiatry, 25*(1), 125.

Menear, M., Girard, A., Dugas, M., Gervais, M., Gilbert, M., & Gagnon, M. P. (2022). Personalized care planning and shared decision making in collaborative care programs for depression and anxiety disorders: A systematic review. *PLoS ONE, 17*(6), Article e0268649.

Mierau, S. B. (2025). Do I Have ADHD? Diagnosis of ADHD in Adulthood and Its Mimics in the Neurology Clinic. *Neurology: Clinical Practice, 15*(1), e200433.

Mongelli, F., Georgakopoulos, P., & Pato, M. T. (2020). Challenges and opportunities to meet the mental health needs of underserved and disenfranchised populations in the United States. *Focus, 18*(1), 16–24.

Nieuwenhuis, S., Nijs, S., & Maes, B. (2025). Person-Centred Care for People with Profound Intellectual and Multiple Disabilities: A Rapid Review. *Journal of Long-Term Care, 2025*, 110–126.

Nowrouzi-Kia, B., Bani-Fatemi, A., Jackson, T. D., Li, A. K. C., Chattu, V. K., Lytvyak, E., Straube, S., et al. (2025). Evaluating the efficacy of telehealth-based treatments for depression in adults: A rapid review and meta-analysis. *Journal of Occupational Rehabilitation, 35*(4), 703–724.

Oliver, M. (2013). The social model of disability: Thirty years on. *Disability & Society, 28*(7), 1024–1026.

Padesky, C. A., & Mooney, K. A. (2012). Strengths-based cognitive–behavioural therapy: A four-step model to build resilience. *Clinical Psychology & Psychotherapy, 19*(4), 283–290.

Parikh, S. V., Quilty, L. C., Ravitz, P., Rosenbluth, M., Pavlova, B., Grigoriadis, S., & CANMAT Depression Work Group et al. (2016). Canadian Network for Mood and Anxiety Treatments (CANMAT) 2016 clinical guidelines for the management of adults with major depressive disorder: section 2. Psychological treatments. *The Canadian Journal of Psychiatry, 61*(9), 524–539.

Penninx, B. W., Milaneschi, Y., Lamers, F., & Vogelzangs, N. (2013). Understanding the somatic consequences of depression: Biological mechanisms and the role of depression symptom profile. *BMC Medicine, 11*(1), 129.

Pinals, D. A., Hovermale, L., Mauch, D., & Anacker, L. (2022). Persons with intellectual and developmental disabilities in the mental health system: Part 1. *Clinical Considerations. Psychiatric Services, 73*(3), 313–320.

Polderman, T. J., Boomsma, D. I., Bartels, M., Verhulst, F. C., & Huizink, A. C. (2010). A systematic review of prospective studies on attention problems and academic achievement. *Acta Psychiatrica Scandinavica, 122*(4), 271–284.

American Psychiatric Association. (2015). *The American Psychiatric Association practice guidelines for the psychiatric evaluation of adults.* American Psychiatric Association.

Quiroz Santos, E., Stein, L. A. R., Stamates, A., & Voyer, H. (2025). The Impact of Peer-Based Recovery Support Services: Mediating Factors of Client Outcomes. *The Journal of Behavioral Health Services & Research*, 1–27.

Radell, M. L., Abo Hamza, E. G., Daghustani, W. H., Perveen, A., & Moustafa, A. A. (2021). The impact of different types of abuse on depression. *Depression Research and Treatment, 2021*(1), 6654503.

Ramar, K., Malhotra, R. K., Carden, K. A., Martin, J. L., Abbasi-Feinberg, F., Aurora, R. N., Trotti, L. M., et al. (2021). Sleep is essential to health: An American Academy of Sleep Medicine position statement. *Journal of Clinical Sleep Medicine, 17*(10), 2115–2119.

Ranjbar, N., Erb, M., Mohammad, O., & Moreno, F. A. (2020). Trauma-informed care and cultural humility in the mental health care of people from minoritized communities. *Focus, 18*(1), 8–15.

Ridley, M., Rao, G., Schilbach, F., & Patel, V. (2020). Poverty, depression, and anxiety: Causal evidence and mechanisms. *Science, 370*(6522), eaay0214.

Rubin, K. H., Coplan, R. J., & Bowker, J. C. (2009). Social withdrawal in childhood. *Annual Review of Psychology, 60*(1), 141–171.

Salvagioni, D. A. J., Melanda, F. N., Mesas, A. E., González, A. D., Gabani, F. L., & De Andrade, S. M. (2017). Physical, psychological and occupational consequences of job burnout: A systematic review of prospective studies. *PLoS ONE, 12*(10), Article e0185781.

Shelton, R. C., & Hollon, S. D. (2012). The long-term management of major depressive disorders. *Focus, 10*(4), 434–441.

Sheppes, G., Suri, G., & Gross, J. J. (2015). Emotion regulation and psychopathology. *Annual Review of Clinical Psychology, 11*(1), 379–405.

Sobin, C. (1997). Psychomotor symptoms of depression. *American Journal of Psychiatry.*

Solomon, D. A., Keller, M. B., Leon, A. C., Mueller, T. I., Lavori, P. W., Shea, M. T., Endicott, J., et al. (2000). Multiple recurrences of major depressive disorder. *American Journal of Psychiatry, 157*(2), 229–233.

Stein, D. J., Shoptaw, S. J., Vigo, D. V., Lund, C., Cuijpers, P., Bantjes, J., Maj, M., et al. (2022). Psychiatric diagnosis and treatment in the 21st century: Paradigm shifts versus incremental integration. *World Psychiatry, 21*(3), 393–414.

Suchy, Y. (2009). Executive functioning: Overview, assessment, and research issues for non-neuropsychologists. *Annals of Behavioral Medicine, 37*(2), 106–116.

Tetzlaff, L., Schmiedek, F., & Brod, G. (2021). Developing personalized education: A dynamic framework. *Educational Psychology Review, 33*(3), 863–882.

Thomas, M. S., Crosby, S., & Vanderhaar, J. (2019). Trauma-informed practices in schools across two decades: An interdisciplinary review of research. *Review of Research in Education, 43*(1), 422–452.

Torous, J., Linardon, J., Goldberg, S. B., Sun, S., Bell, I., Nicholas, J., Firth, J., et al. (2025). The evolving field of digital mental health: Current evidence and implementation issues for smartphone apps, generative artificial intelligence, and virtual reality. *World Psychiatry, 24*(2), 156–174.

Torrey, W. C., Griesemer, I., & Carpenter-Song, E. A. (2017). *Beyond "med management"*.

Troyer, A. K. (2018). Activities of Daily Living (ADL). In *Encyclopedia of Clinical Neuropsychology* (pp. 45–47). Springer.

Vaillancourt, D. E., & Newell, K. M. (2003). Aging and the time and frequency structure of force output variability. *Journal of Applied Physiology, 94*(3), 903–912.

Vos, L. M., Bronstein, M. V., Gendron, M., Joormann, J., & Everaert, J. (2025). Navigating the social world: the interplay between cognitive and socio-affective processes in depression and social anxiety. *Cognitive Therapy and Research*, 1–16.

Wampold, B. E., & Flückiger, C. (2023). The alliance in mental health care: Conceptualization, evidence and clinical applications. *World Psychiatry, 22*(1), 25–41.

Wang, J., Mann, F., Lloyd-Evans, B., Ma, R., & Johnson, S. (2018). Associations between loneliness and perceived social support and outcomes of mental health problems: A systematic review. *BMC Psychiatry, 18*(1), 156.

Williams, A. E., Fossey, E., Corbière, M., Paluch, T., & Harvey, C. (2016). Work participation for people with severe mental illnesses: An integrative review of factors impacting job tenure. *Australian Occupational Therapy Journal, 63*(2), 65–85.

Winstein, C. J., Stein, J., Arena, R., Bates, B., Cherney, L. R., Cramer, S. C., Zorowitz, R. D., et al. (2016). Guidelines for adult stroke rehabilitation and recovery: A guideline for healthcare professionals from the American Heart Association/American Stroke Association. *Stroke, 47*(6), e98–e169.

Woo, Y. S., Rosenblat, J. D., Kakar, R., Bahk, W. M., & McIntyre, R. S. (2016). Cognitive deficits as a mediator of poor occupational function in remitted major depressive disorder patients. *Clinical Psychopharmacology and Neuroscience, 14*(1), 1.

Chapter 2
Neurological Perspectives on Disability in Unipolar Depression

Juveriya Israr, Shabroz Alam, and Ajay Kumar

Abstract Unipolar depression, also known as Major Depressive Disorder (MDD), is a condition that has a profound effect on the everyday functioning of an individual, going beyond just mood changes. Recent studies have emphasized the importance of a number of structural, physiological, neurochemical, neuroendocrine, and inflammatory problems in the nervous system in the exacerbation of cognitive, emotional, and social interaction problems in MDD. It is important to note that problems related to attention, memory, executive function, and motivation are linked to structural problems in the most important regions of the brain, such as the prefrontal cortex, hippocampus, and anterior cingulate cortex. Moreover, problems with the functioning of the brain networks, such as the default mode, central executive, and salience networks, are also common. Neuroinflammation further adds to the problems associated with MDD, as it is also responsible for problems such as fatigue, psychomotor slowing, and social isolation. In addition, neurotransmitter dysregulation and hyperactivity of the hypothalamic–pituitary–adrenal (HPA) axis worsen the functional detriments in those with MDD. This chapter interprets the pathophysiology of unipolar depression at a neurobiological level in reference to the knowledge we have gained using state-of-the-art neuroscience today. It emphasizes the significance of these networks for the development of targeted treatment interventions e.g. (cognitive) remediation, transcranial magnetic stimulation, psychotherapy, and medication. A comprehensive understanding of various neurobiological perspectives would be critical to developing individualized treatment strategies that are specifically tailored for the prevention of disability and restoration of functioning in MDD patients.

J. Israr
Institute of Biosciences and Technology, Shri Ramswaroop Memorial University (SRMU), Barabanki, India

J. Israr · S. Alam
Department of Biotechnology, Era University, Lucknow, India

A. Kumar (✉)
Department of Biotechnology, Rama University, Kanpur, India
e-mail: ajaymtech@gmail.com

A. Kumar (ed.), *Integrating Disability Care in Major Depressive Disorder Through Neurological and Mental Health Perspectives for Empowerment*,
SpringerBriefs in Modern Perspectives on Disability Research,
https://doi.org/10.1007/978-981-95-8955-5_2

Keywords Unipolar depression · Major depressive disorder · Disability · Cognitive dysfunction · Neurobiology · Brain networks · Neurotransmitters · Neuroinflammation · HPA axis

2.1 Introduction: Linking Neuroscience and Disability in Unipolar Depression

Unipolar depression, or Major Depressive Disorder (MDD) is identified as a neurobiological condition, which has a severe impact on such brain regions as cognition, emotional regulation, motivation, and executive functioning. It is marked by severe disabilities in the daily living, socialization, self-care and professional abilities. The recent findings in neuroscience stress the value of the structural and functional alterations in the brain that explain the functional deficits in patients and shifts the emphasis, which was based on classical clinical approaches, where the focus was made on emotional conditions, including sadness and anhedonia (American Psychiatric Association, 2022; World Health Organization, 2017). Current neuroimaging results (Drevets et al., 2008; Malykhin et al., 2010; Pizzacalli & Roberts, 2022) have continuously shown distorted patterns of brain functioning in people with depression by using such techniques as electroencephalography (EEG), positron emission tomography (PET), functional magnetic resonance imaging (fMRI) and studies of cognitive control networks. These neurological alterations are closely related to the cognitive and psychosocial disabilities normally seen in unipolar depression, and, concerningly, these impairments can still be found after the symptoms of mood have subsided, which may lead to long-term disabilities (Lam et al., 2014; McIntyre et al., 2013a, 2013b). Understanding the neurological basis of dysfunction in unipolar depression is vital in informing practice assessment, intervention therapies, rehabilitation approaches and policy making. This chapter combines the main results of neuroinflammation, neuroendocrine dysregulation, cognitive functions, brain network activity, and the general brain structure to outline the cognitive neurological processes that determine the impairment in people affected by unipolar depression.

2.2 Structural Brain Changes Contributing to Disability

Studies in neuroimaging of unipolar depression have shown that there are structural brain abnormalities related to functional impairments. Interesting results of the research carried out by (Pizzagalli & Roberts, 2022) and (Rajkowska, 2000) show that people with depression tend to have a smaller amount of grey matter or cortical thickness in the prefrontal cortex that is part of the planning, decision-making, executive functions, and inhibition processes. Additional studies by (Lam et al., 2014)

Table 2.1 Structural brain abnormalities and functional impairments in unipolar depression

S. No	Brain region	Structural change	Functional impact/disability
1	Dorsolateral Prefrontal Cortex (DLPFC)	Reduced cortical thickness / volume	Impaired executive function, planning, attention, task initiation
2	Hippocampus	Volume loss	Memory deficits, poor learning, stress vulnerability
3	Anterior Cingulate Cortex (ACC)	Reduced volume	Impaired motivation, error detection, emotional regulation
4	Amygdala	Hyperactivity / altered connectivity	Heightened emotional reactivity, social withdrawal

support the fact that cognitive sluggishness, reduced attention, poor decision-making capacity, and problems with task initiation are associated with changes in the dorsolateral prefrontal cortex (DLPFC), which has an adverse impact on job performance and daily life. Besides, repeated or persistent depression is linked to the downward changes in the hippocampal volumes, as demonstrated in the studies by (Malykhin et al., 2010) and (Frodl et al., 2006). The hippocampus is the critical component of memory and learning; its deterioration may result in the inability to study successfully, disorder in the family, and difficulties with work-related problems-solving. (Lam et al., 2014) also mention that it is associated with impairments in episodic memory, slower learning rates, and a reduced ability to withstand stress.

Another prominent part of the brain that is impaired in the case of depression is the anterior cingulate cortex (ACC), which entails emotional regulation, error detection, and enhancement of motivation (Drevets et al., 2008). The decrease in volume of the ACC as shown in (Table 2.1) complicates the ability of individuals to move on the transition between thought and action, hinders the decision-making process, and reduces the level of motivation. These changes in the brain structure reveal the multifaceted interaction between the integrity of the brain and the cognitive-emotional difficulties experienced by people with unipolar depression.

2.3 Functional Brain Network Dysregulation

There is growing momentum of the understanding that unipolar depression is a systemic disorder of brain networks and not merely localized abnormalities. Patients with major depressive disorder (MDD) have severe functional impairment and disability which is aggravated by disturbances in balance and interaction of intrinsic functional brain systems. The networks play a critical role in coordinating cognitive, affective, and attentional activities. It is worth noting that the Default Mode Network (DMN), Central Executive Network (CEN) and Salience Network (SN) are crucial in the pathophysiology of the depressive symptoms. DMN has been linked to self-referential thoughts, CEN is linked to cognitive control, and SN enhances adaptive attentional shifting in people with MDD, and these are critical to the efficient emotional control and cognitive processes (Fig. 2.1).

2.3.1 *Default Mode Network (DMN):*

Associated with self-referential thinking and rumination. The Default Mode Network (DMN), the networks of medial prefrontal cortex, anterior cingulate cortex/precuneus and lateral parietal areas are essential in helping to support inward directed cognition including autobiographical memory and self-referential processing. It is generally less active when people are in the best health condition and are preoccupied with external stimuli. Nonetheless, the same pattern is observed in people with Major Depressive Disorder (MDD); they are hyper connected and hyperactive in the DMN even when engaged in procedures (Hamilton et al., 2011). This process is strongly intertwined with one of the primary symptoms of depression namely a cyclic and inactive obsessive–compulsive focus on negative thoughts and feelings, which is often a consequence of dysadaptive rumination. The heightened connection in the DMN, especially between the medial prefrontal cortex and the posterior cingulate cortex, interferes with the capacity of individuals to stop negative, self-centered ruminations. Consequently, this hyperconnectivity leads to an increase in the risk of occurrence of depressive states. Affected individuals can experience poor concentration, chronic melancholy, and intrusive negative self-assessment, and this will result in a decrease in their psychosocial and occupational functioning.

2.3.2 *Central Executive Network (CEN)*

Governs attention, working memory, and problem-solving. The Central Executive Network (CEN) which is found in the dorsolateral prefrontal cortex (DLPFC) and the posterior parietal cortex is of significant importance in the higher-order cognitive processes including working memory, attention management, planning, and problem-solving. The best possible operation of the CEN enables people to adjust to their surroundings and control their emotions. Yet, under depression conditions, hypoactivation and loss of functional connections in the CEN, and particularly in the DLPFC, are also noticeable as found in the work by (Lam et al., 2014) and (Pizzacalli & Roberts, 2022). This dysfunction leads to a number of cognitive disorders such as indecisiveness, loss of concentration, impaired cognition and loss of goal-directed behavior that ultimately cause cognitive regulation and attentional capacity. The dysfunctional functioning of the CEN further intensifies negative cognitive loops that make it difficult to control rumination caused by the Default Mode Network

(DMN). This is an imbalanced response between introspective thinking and externally oriented cognition which is one of the major neural processes associated with cognitive impairment in Major Depressive Disorder (MDD).

2.3.3 *Salience Network (SN)*

Detects and prioritizes important internal or external stimuli. The Salience Network (SN), which is composed of the anterior cingulate cortex and the insula, is critical in determining behaviorally interesting stimuli and establishing the dynamic interface between the Default Mode Network (DMN) and the Central Executive Network (CEN). This network plays a role in dynamic distribution of cognitive and attentional resources by combining information of physiological, emotional, and cognitive nature. This important switching mechanism is impaired in the case of Major Depressive Disorder (MDD) in people as the SN is not able to coordinate both internal and external processing modes (Menon, 2011). In its turn, a dysfunctional SN results in a greater need to focus on negative emotional signs and experienced distress, instead of utilizing the CEN to perform cognitively demanding activities. This deficiency is defined by inability to adjust to the new environment, a great level of attentiveness, and an overreaction of emotions. Consequently, depressed people often state that they are stuck in negative thoughts and feelings and that they are not able to alter their behavior and create ways to handle their problem. (McIntyre et al., 2013a, 2013b) reported that the extent of functional impairment of MDD is significantly correlated with impaired connectivity between these neural networks, which suggests that a disturbance in the architecture of the brain may prove to be an effective predictor of functional impairment as compared to the severity of the depressive symptoms.

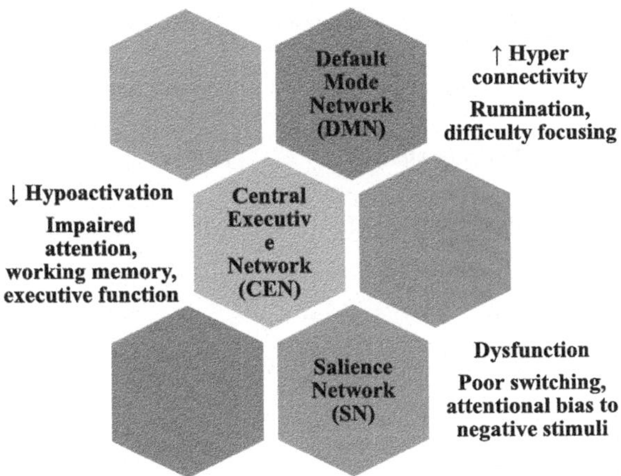

Fig. 2.1 Functional Brain Network Dysregulation in Unipolar Depression

2.4 Cognitive Dysfunction as a Neurological Source of Disability

Specifically, cognitive impairment is especially related to unipolar depression and is a significant factor that leads to disability in individuals since they are not able to carry out their daily activities. Major depressive disorder sufferers usually have a number of cognitive impairments including lack of concentration, loss of processing speed, weak working memory, executive malfunction and high distractibility. The studies show that these cognitive difficulties disrupt sustained attention and make it difficult to organize daily life tasks and/or perform a job (Lam et al., 2014). A slow processing speed means that it is not possible to perform routine tasks on time, which makes tasks less efficient (McIntyre et al., 2013a, 2013b). Additionally, there is executive dysfunction which is a barrier to planning and making decisions that reduces the capacity to perform complex and goal-oriented activities and adapt new circumstances (Clark & Beck, 2010). Also, impairments of the working memory reduce the ability to temporarily hold and process information affecting learning and problem-solving skills that are instrumental to academic and professional achievement. It is important to note that such cognitive impairments do not disappear but instead occur with mood symptoms, which supports their status as distinct indicators of occupational impairment. This observation is supported by neurological evidence (Lam et al., 2014; Porter et al., 2013), which is also accompanied by such a phenomenon as cognitive residue, when cognitive impairment persists despite mood improvement. This highlights the importance of treating and evaluating cognitive dysfunction in depression, and not only concerning the reduction of mood symptoms (McIntyre et al., 2013a, 2013b) (Table 2.2).

Table 2.2 Cognitive domains affected in unipolar depression and related disability

S. No	Cognitive domain	Neurological basis	Functional impairment
1	Attention & Concentration	DLPFC, parietal cortex	Errors at work, poor task completion
2	Executive Function	DLPFC, ACC	Difficulty planning, problem-solving
3	Working Memory	DLPFC, hippocampus	Impaired multitasking, daily task management
4	Processing Speed	White matter connectivity	Slowed daily routines, reduced productivity

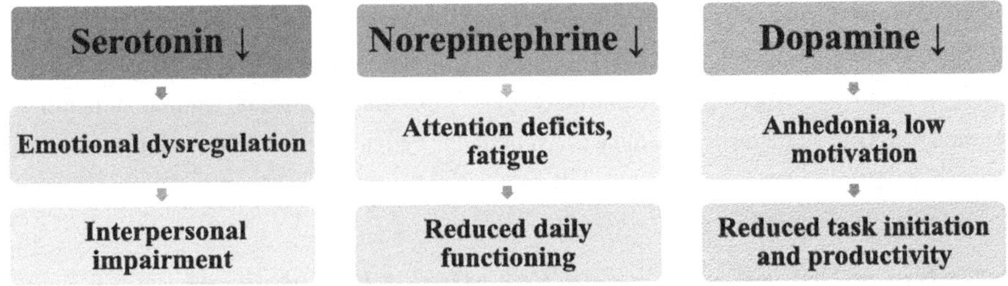

Fig. 2.2 Neurotransmitter dysregulation and functional disability

2.5 Neurotransmitter Systems and Functional Impairment

Depression is also linked to the dysregulation of a number of neurotransmitter systems, especially those of serotonin, norepinephrine, and dopamine. These faulty systems cause adverse outcomes in attention, motivation, and level of social interaction which are fundamental aspects of ordinary functioning. Research conducted by (Whitton et al., 2015), the (American Psychiatric Association, 2022) and (Clark & Beck, 2010) indicates that patients with the Major Depressive Disorder (MDD) are normally challenged by lack of ability to engage in goal-oriented behaviour and reduced ability to start things. It is also worth noting that these challenges are associated with impairments of the dopaminergic system. The interaction of changes in neurotransmitters, structural changes, and the changes in brain network, results in the development of a broad range of functional deficits in the depressed individual (Fig. 2.2).

2.6 Neuroendocrine Mechanisms: HPA Axis Dysregulation

One of the major neuroendocrine pathologies, that is related to the presence of unipolar depression, is the hyperactivity of the hypothalamic–pituitary–adrenal (HPA) axis which causes functional impairment. (Menke, 2024) and (Whitton et al., 2015) report that depressed people tend to have a low ability to regulate stress responses, which is characterized by high levels of basal cortisol, low levels of stress responsiveness, disturbed circadian rhythms, and weak negative feedback inhibition of the HPA axis. The chronic increase in cortisol may result in damaging alterations in the brain structure and functioning, which include atrophy of the essential brain areas, including the hippocampus, which is related to the inability to consolidate memory, reduced neurogenesis, and increased susceptibility to anxiety. Also, sleep disorders of hypersomnia and insomnia can be caused by the disruption of the circadian rhythms in the body. Such upheavals worsen the already elevated deficits caused by the depression factor, and cause cognitive impairments, emotional insecurity, fatigue, and difficulties in day-to-day functioning (Harvey, 2011).

Table 2.3 Inflammatory mechanisms contributing to disability

S. No	Marker	Neurological effect	Functional impact
1	IL-6	Alters synaptic plasticity	Cognitive slowing, learning difficulties
2	TNF-α	Reduces dopamine transmission	Fatigue, reduced motivation
3	CRP	Induces neuroinflammation	Psychomotor slowing, low activity

2.7 Neuroinflammation and Functional Impairment

According to a growing body of research, there is a high correlation between depression without any manic episodes and raised markers of inflammation such as interleukin-6 (IL-6), tumor necrosis factor–alpha (TNF-α) and C-reactive protein (CRP). This provides a potential biological explanation for people with unipolar depression as well as indicating that disturbances in inflammation are therefore likely to be relevant to the pathophysiology of major depressive disorder (MDD). Consistent with this view, individuals with MDD have significant functionality limitations that are associated with increased levels of inflammation (Lee & Giuliani, 2019; Miller & Raison, 2016). Table 2.3 illustrates that individuals with unipolar depression have elevated levels of IL-6, TNF-α and CRP compared to non-depressed individuals. Fatigue and psychomotor slowing down are two major disabling indicators of MDD and appear to be directly related to inflammation rather than just related to emotional disturbances. In addition, heightened levels of inflammation have been associated with an increase in the occurrence of anhedonia, social withdrawal, and a decrease in physical activity. This helps to highlight the extent to which biological and psychological factors are related through two way causal processes in individuals with MDD.

2.8 Neurological Basis of Emotional Dysregulation and Social Disability

Changes within the limbic-prefrontal neural circuits correlate strongly with difficulties in the regulation of emotion that occur in people with unipolar depression. These people have amygdala hyperactivity and a lack of activity in the prefrontal regions that mediate top-down regulation. This imbalance creates a greater degree of reactivity to negative emotions and a reduced ability to effectively regulate emotional responses (Drevets et al., 2008; Whitton et al., 2015). Thus, these neurological deficits can create a number of functional problems such as impaired decision-making in social situations; reduced ability to cope with stress; withdrawal from social situations; and increased conflict with others. In addition, the insula and the anterior cingulate cortex (ACC), two critical regions for emotional intelligence, empathy, and the processing of emotion, are profoundly altered in people who suffer from unipolar depression. Interpersonal and psychosocial deficits are frequently reported by those who suffer from this illness. The failure to effectively manage these deficits will lead to further exacerbation of the condition by decreasing social participation and complicating interpersonal relationships (Menon, 2011). The complex interplay of neurological alterations and psychosocial outcomes illuminates both the multi-dimensional nature of unipolar depression and the effects on emotional and social functioning.

2.9 Neurological Predictors of Functional Outcomes

Recent studies in neuroscience show that severity of symptoms on their own does not offer a full understanding of the real-life impairments experienced by people diagnosed with unipolar depression; these studies suggest that there is a need to identify neurobiological markers as predictors of social, vocational, and daily functioning compared to clinical assessments based on mood. An important finding is diminished levels of dorsolateral prefrontal cortex (DLPFC) activity associated with functional impairment. The DLPFC is a key component of the Central Executive Network, and decreased levels of activity in this brain region are correlated with reduced ability to perform in many areas of daily life (i.e., attention, working memory, executive function/decision-making, productivity and employment) (Lam et al., 2014). Furthermore, the DLPFC will continue to show decreased functioning even if mood symptoms have partially or completely resolved, which indicates that decreased DLPFC activity is an enduring marker of cognitive and functional deficits. Hippocampal volume is a substantial brain volume marker for outcome in depression and was found to be an indicator of long term functioning in both chronic or recurrent depression. Decrease in the size of the hippocampus has been associated with chronic states of depression, affecting both memory consolidation and regulation of stress with both resulting in challenges to day to day functioning and increased complexity of managing stress, resulting in greater functional deficits over time (Frodl et al., 2006). Another negative consequence of hyperconnectivity in the Default Mode Network is that it strongly correlates to increased levels of rumination and therefore correlates to decreased levels of engagement in each of the activity that a person is attempting. Excessively hyperconnected networks inhibit movement or impeded transitions from cognitions that are self-referential (i.e. passage of time) to activity-based (i.e. goal completion). As these shifts do not occur cognitive inefficiencies occur and will continue to lead to greater cognitive disconnection and functional disengagement in day to day activities. (Hamilton et al., 2011).

The role of peripheral inflammatory markers will be important for understanding various functional impairments present in depressed people. For example, elevated pro-inflammatory cytokine levels have been associated with various depression-related functional impairments such as fatigue, lack of motivation, psychomotor retardation, and disruptions in the ability to carry out many instrumental daily living (IADL) tasks like financial responsibility and household chores (Lee & Giuliani, 2019). Also, symptoms from chronic inflammation typically do not respond to commonly used antidepressant medications indicating an area in need of further study. Overall, these data indicate that inflammatory processes, structural and functional brain changes, and specific neurobiological markers play integral roles in determining the long-term functional prognosis of people with unipolar depression. These biomarkers also tend to better predict functional outcomes than simply using emotional symptoms severity alone as predictors which supports the need for implementing a disability-based approach to assessment and intervention. To maximize recovery and improve long-term outcomes for depressed individuals requires an understanding of cognitive processes, stabilization of neural networks and examination of molecular pathways resulting in functional impairment (Fig. 2.3).

2.9.1 Implications for Treatment and Rehabilitation

Learning about how the brain works in the person with unipolar depression is key to making decisions about creating treatment strategies and optimizing functional rehabilitation. Current treatment modalities in this patient population tend to emphasize cognitive methods, dysfunctional neural networks, and functional performance rather than simply targets for alleviating mood symptoms due to increasing evidence supporting the efficacy of targeted, mechanism-based treatments that both improve symptoms and provide support for functional recovery. For example, cognitive remediation therapy (CRT) focuses on cognitive deficits that are common among individuals suffering from depression and typically impact areas such as attention, working memory and executive functions within an individual's cognitive functioning abilities. The CRT facilitates the long-term reorganization of neural pathways related to goal directed behavior and problem-solving by using systematic cognitive exercises with the individual to produce measurable improvement in an individual's ability to function in occupational settings, educational settings and within everyday living for individuals who suffer from chronic mood disorders. Neuromodulation methods like TMS directly target areas of imbalance in the brain's circuitry. rTMS applied to the DLPFC has been shown to improve emotion regulation and cognitive control, as well as executive function, by creating synchronization between the limbic and default mode networks and enhancing the activity of the Central Executive Network. Additionally, these treatments provide considerable functional improvements, such as reducing cognitive impairments and improving occupational functioning, in addition to their antidepressant effects.

In addition, rehabilitative techniques are largely based on lifestyle-based interventions. For example, it has been shown that regular behavioral activation and physical activity increase neuroplasticity, reduce systemic inflammation, and increase the level of neurotrophic factors like brain-derived neurotrophic factor (BDNF). Exercise also acts as an effective adjunct treatment for depression, improving cognition, energy levels, motivation and resilience to stress. Psychotherapeutic techniques have been shown to promote functional recovery. Cognitive Behavioral Therapy is particularly effective in doing so as it affects the prefrontal-limbic pathways which control cognitive functioning and emotions. The application of CBT has been found to reduce

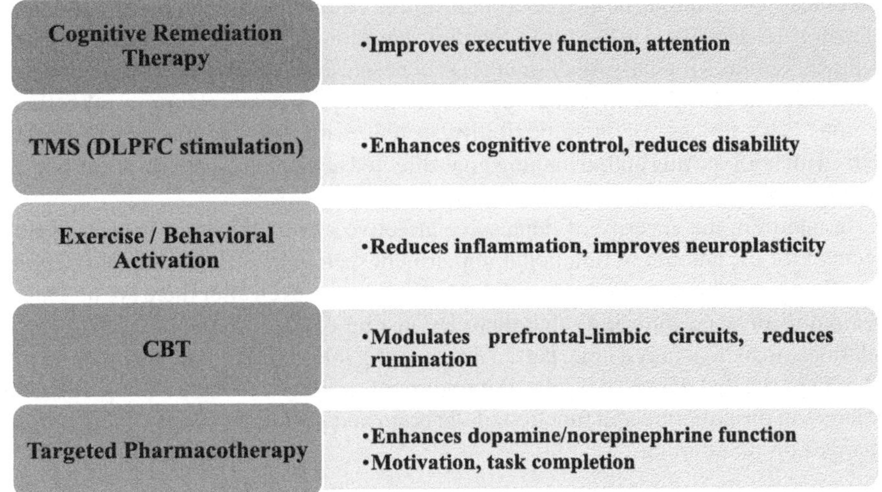

Fig. 2.3 Neurology-Informed Interventions for Functional Disability

hyperactivity in the Default Mode Network and improve regulation of emotions by the prefrontal cortex. The use of CBT has subsequently improved social functioning, decision-making abilities and the management of stress via targeting cognitive distortions, behavioral avoidance and ruminative thought patterns. Pharmaceutical strategies should also pivot toward alleviating functional impairments rather than merely addressing mood symptoms. Medications that enhance dopaminergic or noradrenergic transmission can provide substantial benefits for patients looking to boost motivation, cognitive processing speed, and executive functions elements crucial to their daily and professional performance. Particularly, individuals with chronic cognitive impairments, psychomotor retardation, or anhedonia stand to gain significantly from these pharmaceutical options, which can lead to marked reductions in disability. To summarize, all types of therapies for unipolar depression point to the need to take a holistic, neuroscience-based approach to treatment in order to maximize recovery and enhance quality of life over the long term by addressing cognitive dysfunction, inflammation, neural network dysregulation, and motivational deficits directly. The combination of these cutting-edge techniques with typical care can help to close the gap between conventional psychosocial rehabilitation and effective relief from depressive symptoms.

2.9.2 *Brain–Behavior Pathways to Functional Disability*

Functional ability is affected by neurological irregularities due to unipolar depression through the breakdown of the connections between brain function and behavior.

Specifically, alterations in the pre-frontal-limbic circuit greatly affect the regulation of major areas involved in executive function, emotional regulation, and goal-directed behavior required for daily functioning (Disner et al., 2011; Mayberg, 1997). A significant decrease in top-down regulatory control from prefrontal regions to subcortical regions leads to less cognitive flexibility, trouble with decision making, and problems with both starting and sustaining goal-directed behaviors (Arnsten, 2009; Kaiser et al., 2015).

In addition, the severity of depressive affective symptoms does not completely account for the amount of functional impairment seen in those with unipolar depression. For example, neurobiological deficits may remain even after there are no longer symptoms present, indicating that there are lasting changes in the brain-behavioral relationship, which play a large part in determining long-term functional impairment. This suggests that depressive-related impairment should be thought of as permanent changes in the structure and function of the brain and not simply as an effect of mood changes on functioning.

2.9.3 Neuroplasticity and the Progression of Functional Impairment

Recurrent and chronic unipolar depression appears to create maladaptive neuroplastic changes through the accumulation of functional decline; When there is an extended period of time experiencing prolonged stress in your life, or combined with elevated levels of glucocorticoids, these will negatively affect dendritic structure and synaptic density as well as decrease neurogenesis, particularly in the hippocampus and prefrontal cortex. All of these structural changes from the previous statements may be linked to some degree to deficits in learning, memory and executive functioning which will serve to reinforce behavioral patterns that will hinder one's recovery. Additionally, the theory of neuroprogression indicates that repeated episodes of unipolar depression generate cumulative experience related to the brain over a period of time, resulting in increased risk for future episodes with associated declines in functioning compared to the level achieved prior to that episode. This cumulative experience may partially explain individuals' persistent level of dysfunction despite adequate management of depressive symptoms. It is therefore critical to further define unipolar depression as a disorder characterized by dysfunctional and impaired neuroplasticity, and thus it is necessary to provide early and ongoing intervention(s) addressing the preservation of neural health and functionality.

2.9.4 Psychosocial and Occupational Implications of Neurological Dysfunction

Neurologic impairments caused by major depression result in serious consequences for both your mental health and job performance. Neural pathways involved in motivation, rewards and cognitive control are disrupted due to these neurologic deficits, which lead to decreased productivity, increased absenteeism from work, and difficulty maintaining employment (Lerner & Henke, 2008; Treadway & Zald, 2011). In addition, cognitive deficits such as poor processing speed; poor attention; and poor working memory add to the overall difficulty of meeting job performance expectations and adapting to high-demand work environments (Evans et al., 2014; McIntyre et al., 2013a, 2013b). In addition, social functioning is negatively affected by neurological disorders as they alter prefrontal and limbic areas of the brain's activity leading to abnormal functioning of social cognition (or how we think about other people). This impacts our ability to perceive emotions in others and hold the ability for empathy therefore making it less likely that we will relate appropriately to others causing social isolation and relationship problems as noted by (Kupferberg et al., 2016; Weightman et al., 2014). Social impairments can result in fewer opportunities to have supportive social relationships, which increases the extent of functional disability and may also contribute to a more rapid progression of the disorder.

2.9.5 Methodological Challenges and Future Research Directions

Even though there is a lot of improvement in neuroimaging technology and cognitive neuroscience; there are still major methodological problems to be overcome when trying to learn about what connects (or relates) areas of the brain that are malfunctioning due to having a unipolar depression and how this connects with an individual's disability level. There are many inconsistencies within the existing body of literature due to many variables influencing the results including the length of time that an individual has had the illness, how often they have experienced episodic depressed states, their history of receiving treatment and whether or not they have comorbid conditions as reviewed by; (Fried & Nesse, 2015) and (Dichter et al., 2015). Furthermore, researchers' reliance on cross-sectional research methods will severely limit their ability to make any causal statements regarding the relationship between brain pathology and functional impairment due to the point-in-time nature of the data. To address this limitation, future research must include longitudinal and multimodal approaches that incorporate neuroimaging methods, cognitive assessment tools, inflammatory markers and ecological outcome measures (Insel et al., 2010; McTeague et al., 2016). Once researchers can identify neurological markers for functional recovery, rather than just concentrating on symptom relief, they will have established a foundation for developing individualized treatments based on neuroscientific evidence. Developing effective ways to provide these potential treatments should eventually help transition patients from temporary to sustained increases in self-reported quality of life and functional ability.

2.10 Conclusion

The main issue with unipolar depression is the neurological impairment as shown through a wide variety of means including differences in structure of the brain, differences in the neural networks, compared to "normal" people, compared to how they used to be, resulting in an imbalance between their neurotransmitters, and problems with the HPA axis and neuroinflammatory activity, and these issues lead to all of the cognitive impairment, social withdrawal, decreased performance on the job and difficulties with daily living that are common with depression. Therefore, treatment needs to be focused on functional recovery rather than just addressing mood symptoms because neurological symptoms may continue even after mood symptoms have improved. The treatment approaches that can provide the best therapeutic effects will be those that use cognitive remediation, neuromodulation, psychotherapy, and pharmacotherapy targeted at improving attention, executive function, motivation, and emotional regulation through incorporating the neurobiological principles into their approaches. Future research should focus on developing innovative treatment strategies for the functional impairments resulting from unipolar depression and understanding the brain mechanisms responsible for causing the disabilities related to unipolar depression.

References

Arnsten, A. F. T. (2009). Stress signalling pathways that impair prefrontal cortex structure and function. *Nature Reviews Neuroscience, 10*(6), 410–422. https://doi.org/10.1038/nrn2648

Berk, M., Dean, O. M., Cotton, S. M., Gama, C. S., Kapczinski, F., Fernandes, B., Malhi, G. S., et al. (2012). Maintenance N-acetyl cysteine treatment for bipolar disorder: A double-blind randomized placebo controlled trial. *BMC Medicine, 10*(1), 91.

Clark, D. A., & Beck, A. T. (2010). Cognitive theory and therapy of anxiety and depression: Convergence with neurobiological findings. *Trends in Cognitive Sciences, 14*(9), 418–424.

Dichter, G. S., Gibbs, D., & Smoski, M. J. (2015). A systematic review of relations between resting-state functional-MRI and treatment response in major depressive disorder. *Journal of Affective Disorders, 172*, 8–17. https://doi.org/10.1016/j.jad.2014.09.028

Disner, S. G., Beevers, C. G., Haigh, E. A. P., & Beck, A. T. (2011). Neural mechanisms of the cognitive model of depression. *Nature Reviews Neuroscience, 12*(8), 467–477. https://doi.org/10.1038/nrn3027

Drevets, W. C., Price, J. L., & Furey, M. L. (2008). Brain structural and functional abnormalities in mood disorders: Implications for neurocircuitry models of depression. *Brain Structure and Function, 213*(1), 93–118.

Duman, R. S., & Aghajanian, G. K. (2012). Synaptic dysfunction in depression: Potential therapeutic targets. *Science, 338*(6103), 68–72. https://doi.org/10.1126/science.1222939

Evans, V. C., Iverson, G. L., Yatham, L. N., & Lam, R. W. (2014). The relationship between neurocognitive and psychosocial functioning in major depressive disorder: A systematic review. *Journal of Clinical Psychiatry, 75*(12), 1359–1370. https://doi.org/10.4088/JCP.13r08939

Fried, E. I., & Nesse, R. M. (2015). Depression is not a consistent syndrome: An investigation of unique symptom patterns in the STAR*D study. *Journal of Affective Disorders, 172*, 96–102. https://doi.org/10.1016/j.jad.2014.10.010

Frodl, T., Schaub, A., Banac, S., Charypar, M., Jäger, M., Kümmler, P., Meisenzahl, E. M., et al. (2006). Reduced hippocampal volume correlates with executive dysfunctioning in major depression. *Journal of Psychiatry and Neuroscience, 31*(5), 316–323.

Hamilton, J. P., Furman, D. J., Chang, C., Thomason, M. E., Dennis, E., & Gotlib, I. H. (2011). Default-mode and task-positive network activity in major depressive disorder: Implications for adaptive and maladaptive rumination. *Biological Psychiatry, 70*(4), 327–333.

Harvey, A. G. (2011). Sleep and circadian functioning: Critical mechanisms in the mood disorders? *Annual Review of Clinical Psychology, 7*(1), 297–319.

Hasselbalch, B. J., Knorr, U., & Kessing, L. V. (2011). Cognitive impairment in the remitted state of unipolar depressive disorder: A systematic review. *Journal of Affective Disorders, 134*(1–3), 20–31. https://doi.org/10.1016/j.jad.2010.11.011

Idlett-Ali, S. L., Salazar, C. A., Bell, M. S., Short, E. B., & Rowland, N. C. (2023). Neuromodulation for treatment-resistant depression: Functional network targets contributing to antidepressive outcomes. *Frontiers in Human Neuroscience, 17*, 1125074.

Insel, T. R., Cuthbert, B. N., Garvey, M., Heinssen, R., Pine, D. S., Quinn, K. J., Sanislow, C. A., & Wang, P. S. (2010). Research domain criteria (RDoC): Toward a new classification framework for research on mental disorders. *American Journal of Psychiatry, 167*(7), 748–751. https://doi.org/10.1176/appi.ajp.2010.09091379

Kaiser, R. H., Andrews-Hanna, J. R., Wager, T. D., & Pizzagalli, D. A. (2015). Large-scale network dysfunction in major depressive disorder: A meta-analysis of resting-state functional connectivity. *JAMA Psychiatry, 72*(6), 603–611. https://doi.org/10.1001/jamapsychiatry.2015.0071

Kupferberg, A., Bicks, L., & Hasler, G. (2016). Social functioning in major depressive disorder. *Neuroscience & Biobehavioral Reviews, 69*, 313–332. https://doi.org/10.1016/j.neubiorev.2016.07.002

Lam, R. W., Kennedy, S. H., McIntyre, R. S., & Khullar, A. (2014). Cognitive dysfunction in major depressive disorder: Effects on psychosocial functioning and implications for treatment. *The Canadian Journal of Psychiatry, 59*(12), 649–654.

Lee, C. H., & Giuliani, F. (2019). The role of inflammation in depression and fatigue. *Frontiers in Immunology, 10*, 1696.

Lerner, D., & Henke, R. M. (2008). What does research tell us about depression, job performance, and work productivity? *Journal of Occupational and Environmental Medicine, 50*(4), 401–410. https://doi.org/10.1097/JOM.0b013e31816bae50

Maes, M., Berk, M., Goehler, L., Song, C., Anderson, G., Gałecki, P., & Leonard, B. (2012). Depression and sickness behavior are Janus-faced responses to shared inflammatory pathways. *BMC Medicine, 10*(1), 66.

Malykhin, N. V., Carter, R., Seres, P., & Coupland, N. J. (2010). Structural changes in the hippocampus in major depressive disorder: Contributions of disease and treatment. *Journal of Psychiatry and Neuroscience, 35*(5), 337–343.

Mayberg, H. S. (1997). Limbic-cortical dysregulation: A proposed model of depression. *Journal of Neuropsychiatry and Clinical Neurosciences, 9*(3), 471–481. https://doi.org/10.1176/jnp.9.3.471

McEwen, B. S. (2004). Protection and damage from acute and chronic stress: Allostasis and allostatic overload and relevance to the pathophysiology of psychiatric disorders. *Annals of the New York Academy of Sciences, 1032*(1), 1–7. https://doi.org/10.1196/annals.1314.001

McIntyre, R. S., Cha, D. S., Soczynska, J. K., Woldeyohannes, H. O., Gallaugher, L. A., Kudlow, P., Baskaran, A., et al. (2013a). Cognitive deficits and functional outcomes in major depressive disorder: Determinants, substrates, and treatment interventions. *Depression and Anxiety, 30*(6), 515–527.

McIntyre, R. S., Cha, D. S., Soczynska, J. K., Woldeyohannes, H. O., Gallaugher, L. A., Kudlow, P., Alsuwaidan, M., & Baskaran, A. (2013b). Cognitive deficits and functional outcomes in major depressive disorder: Determinants, substrates, and treatment interventions. *Depression and Anxiety, 30*(6), 515–527. https://doi.org/10.1002/da.22063

McTeague, L. M., Goodkind, M. S., & Etkin, A. (2016). Transdiagnostic impairment of cognitive control in mental illness. *Journal of Psychiatric Research, 83*, 37–46.

Menke, A. (2024). The HPA axis as target for depression. *Current Neuropharmacology, 22*(5), 904–915.

Menon, V. (2011). Large-scale brain networks and psychopathology: A unifying triple network model. *Trends in Cognitive Sciences, 15*(10), 483–506.

Miller, A. H., & Raison, C. L. (2016). The role of inflammation in depression: From evolutionary imperative to modern treatment target. *Nature Reviews Immunology, 16*(1), 22–34.

Pizzagalli, D. A., & Roberts, A. C. (2022). Prefrontal cortex and depression. *Neuropsychopharmacology: Official Publication of the American College of Neuropsychopharmacology, 47*(1), 225–246.

Porter, R. J., Bourke, C., & Gallagher, P. (2007). Neuropsychological impairment in major depression: Its nature, origin and clinical significance. *Australian & New Zealand Journal of Psychiatry, 41*(2), 115–128. https://doi.org/10.1080/00048670601109881

Porter, R. J., Bowie, C. R., Jordan, J., & Malhi, G. S. (2013). Cognitive remediation as a treatment for major depression: A rationale, review of evidence and recommendations for future research. *Australian & New Zealand Journal of Psychiatry, 47*(12), 1165–1175.

Rajkowska, G. (2000). Postmortem studies in mood disorders indicate altered numbers of neurons and glial cells. *Biological Psychiatry, 48*(8), 766–777.

Rock, P. L., Roiser, J. P., Riedel, W. J., & Blackwell, A. D. (2014). Cognitive impairment in depression: A systematic review and meta-analysis. *Psychological Medicine, 44*(10), 2029–2040. https://doi.org/10.1017/S0033291713002535

Sheline, Y. I., Sanghavi, M., Mintun, M. A., & Gado, M. H. (2013). Depression duration but not age predicts hippocampal volume loss in medically healthy women with recurrent major depression. In *Depression* (pp. 254–263). Routledge.

American Psychiatric Association. (2022). Diagnostic and statistical manual of mental disorders (5th ed., text revision (DSM-5-TR)). American Psychiatric Publishing.

Treadway, M. T., & Zald, D. H. (2011). Reconsidering anhedonia in depression: Lessons from translational neuroscience. *Neuroscience & Biobehavioral Reviews, 35*(3), 537–555. https://doi.org/10.1016/j.neubiorev.2010.06.006

Weightman, M. J., Air, T. M., & Baune, B. T. (2014). A review of the role of social cognition in major depressive disorder. *Frontiers in Psychiatry, 5* (179). https://doi.org/10.3389/fpsyt.2014.00179

Whitton, A. E., Treadway, M. T., & Pizzagalli, D. A. (2015). Reward processing dysfunction in major depression, bipolar disorder and schizophrenia. *Current Opinion in Psychiatry, 28*(1), 7–12.

World Health Organization. (2017). *Depression and other common mental disorders: Global health estimates*. WHO.

Chapter 3
Mental Health Approaches to Disability in Major Depressive Disorder

Laxmi Singh and Ajay Kumar

Abstract Major Depressive Disorder (MDD) remains one of the current major sources of disabilities worldwide, accounting for considerable occupational, social, as well as cognitive, dysfunctions independent of depressive symptoms. Conventional management approaches focusing on reducing depressive symptoms have few benefits in managing disabilities in such patients. The chapter reviews current mental health approaches for managing disabilities in MDD, with emphasis on the major conceptual divide that exists between symptomatic and functional restoration. The chapter analyses the complexities associated with managing disabilities in MDD, which exist in terms of cognitive, psychosocial, as well as occupational, disabilities, besides neurobiological and environmental issues. Current evidence-based approaches in mental health, such as medication, psychological interventions, psychosocial interventions, online platforms for mental health delivery, and community-based interventions, are critically analyzed for their benefits in managing disabilities. Holistic approaches toward comprehensive healthcare delivery programs based on benefits associated with outcome measures for depression, focusing more on benefits associated with social rehabilitation, are explained. The chapter concludes with new approaches for managing disabilities in depression.

Keywords Major depressive disorder · Treatment · Pharmacological · Healthcare · Rehabilitation · Disability

L. Singh · A. Kumar (✉)
Department of Biotechnology, Rama University, Kanpur, India
e-mail: ajaymtech@gmail.com

A. Kumar (ed.), *Integrating Disability Care in Major Depressive Disorder Through Neurological and Mental Health Perspectives for Empowerment*,
SpringerBriefs in Modern Perspectives on Disability Research,
https://doi.org/10.1007/978-981-95-8955-5_3

3.1 Introduction

Major depressive disorder is considered to be amongst the most prevalent and debilitating psychiatric conditions worldwide, with widespread impacts for affected individuals, their families, as well as for the whole of society (Marx et al., 2023). Moreover, apart from the established high lifetime prevalence, MDD has also turned out to be amongst the top causes of YLDs (Years lived with disability) in the current estimation of disease burden (Yan et al., 2024). Notably, as contrasted to most chronic physical conditions, the type of disability caused due to depression persists even if there has been symptomatic improvement, thereby affecting an individual's work, social interactions, as well as their daily life activities. Moreover, the associated costs due to this disability, including loss of productivity, higher utilization of health services, as well as social marginalization over their whole lifetime, turn out to be quite severe. Therefore, major depressive disorder is defined to be not just a serious health issue but also an important issue requiring major interventions at the public health level. Earlier, the evaluation and management of MDD have been primarily based on symptom-targeting approaches, where treatment success was largely measured by the alleviation of depressive symptoms such as feelings of sadness, anhedonia, changes in sleep and appetite. Symptom severity scales remain the cornerstone of treatment outcomes in clinical settings and trials, where treatment is often based on changes in these scales. Although a symptom-targeting treatment strategy has brought about standardized approaches to depression treatment, it also has its caveats. Many patients meeting criteria for symptomatic remission remain disabled in their actual functioning in many aspects, such as cognitive dysfunction, a lack of motivation, reluctance to engage in social interactions, and poor occupational functioning (Table 3.1) (Sağbaş et al., 2025). The presence of persistent disabilities despite the presence of symptom improvement points toward the failure of a strictly clinical approach toward the disorder, which misses the realities of life lived by patients of depression. The strictly symptom-related approach tends to neglect the psychosocial issues associated with the disorder and the effects of work and other psychosocial factors associated with a person's life, whereas cognitive and motivational changes, major constituents of MDD, do not always have a direct correspondence to the mood-related changes and might have different responses to the traditional therapies and remedies available for the disorder.

To address these challenges, there is an increasing appreciation of the importance of function-oriented and recovery-oriented care for people with MDD. Function-oriented care focuses on the rehabilitation of important roles and activities such as employment, education, social participation, and living independently. A concept of recovery-oriented care implies a shift of emphasis from the mere management of depressive symptoms to a wider concept of health that considers autonomy, purpose of life, and quality of health (Derblom et al., 2025). This viewpoint is compatible with the current concept of disability, as represented by the International Classification of Functioning, Disability and Health (ICF) of the World Health Organization, and considers the concept of disability as the intersection between health

Table 3.1 Domains of disability in major depressive disorder (MDD)

Domain	Key impairments	Functional consequences	Common assessment tools
Occupational	Reduced motivation, impaired concentration, slowed decision-making	Absenteeism, presenteeism, job loss, reduced productivity	Sheehan Disability Scale, WSAS
Social	Social withdrawal, impaired communication, reduced emotional reciprocity	Relationship strain, isolation, reduced social support	WHODAS 2.0, Social Adjustment Scale
Cognitive	Deficits in attention, memory, executive function	Impaired learning, poor task execution, reduced independence	Cognitive batteries, DSST
Emotional	Anhedonia, low self-esteem, emotional dysregulation	Reduced engagement in daily activities	Clinical interviews
Daily Living	Fatigue, psychomotor slowing	Difficulty with self-care and routine tasks	WHODAS 2.0

and factors that influence health and functioning instead of simply considering disabilities as the consequence of health problems (Rao et al., 2025). In addition to considering all the above aspects of depression and advocating the use of multi-interventional approaches that consider psychological treatments and psychosocial rehabilitation of patients with depression in addition to pharmacologic management and computer-mediated intervention and community care programs. This chapter continues to develop this paradigm shift and explore how responding to mental health and disability in Major Depressive Disorder can contribute to this shift. It focuses on the problem of the burden of disability represented by Major Depressive Disorder and challenges the current constructs of models of care based on symptoms of depression and instead advocates a model of care that focuses on functional outcomes.

3.2 Conceptualizing Disability in Major Depressive Disorder

Disability in Major Depressive Disorder (MDD) encompasses more than the manifestation of depressive symptoms and reflects an intricate impairment of an affected person's capacity to function and interact with daily life as he or she needs and as society and culture depend on them to do as a basic function of health and well-being. Although sadness, anhedonia, and intellectual changes are fundamental diagnostic characteristics of depression, the functional impairments and effects of depression experienced as a result of roles at work and social and personal functions of self-care are mainly responsible for the human and/or societal and financial toll of this disorder (Hagan et al., 2025).

3.2.1 WHO Definition of Disability: The ICF Framework

The World Health Organization (WHO) also defines disability using the International Classification of Functioning, Disability, and Health (ICF). The ICF uses the biopsychosocial model in defining disability. Per the ICF model, disability can be generally referred to as an umbrella term for impairments, activity limitations, and participation restrictions caused by interaction effects between health factors and contextual factors (Zhang & Wang, 2025). In this perspective, it can be generally referred to that Major Depressive Disorders (MDD) are not only portrayed as some form of mental illness but something that should be perceived as creating body function impairment (e.g., regulating emotional responses, cognitive function impairment), it also restricts activities (e.g., concentration capability, performing tasks), as well as participant capacity (e.g., employment capacity, social interaction capability) (Fig. 3.1). It should be clearly noted in this matter that the ICF also focuses largely on factors like stigma, workplace support, as well as coping as environmental and personal factors in influencing the degree of disability in individuals suffering from depression (Almakrob & Alduais, 2025).

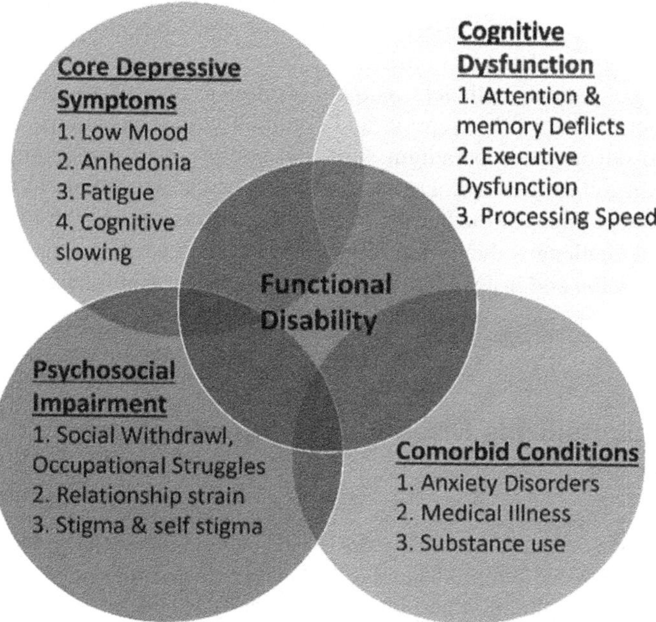

Fig. 3.1 Multidimensional nature of disability in MDD, highlighting interactions between core depressive symptoms, cognitive dysfunction, psychosocial impairment, and comorbid conditions

3.2.2 *Disability, Impairment, and Handicap: Distinctions and Relevance*

Understanding the differences between impairment, disability, and handicap is important for describing functional outcomes in MDD. Impairment indicates the presence of abnormalities in psychological or physiological processes, such as persistent low mood, impaired concentration, or psychomotor retardation (Subedi, 2024). In MDD, they are normally assessed by using scales for symptom severity. Disability reflects the extent to which the sufferer is affected by their inability to do the activities or roles considered normal for everyday life. These may include earning less and interacting less with people if someone develops disability due to their condition and even inability to work if someone has a disability due to lack of cognitive functioning skills or emotional involvement with others as may result from cognitive or emotional changes due to depression. Handicap is a concept initially used as models for understanding disease and less commonly in the concept and framework used by ICF but constitutes social disadvantage which may result from disability and may include social isolation or even expulsion from work due to disability as may exist due to depression as a disease. In MDD, impairments may fluctuate; however, disability may continue even after improvement from symptoms.

3.2.3 *Functional Impairment in Major Depressive Disorder*

3.2.3.1 Work Productivity

Work disability, also referred to as disability related to work, has proven to be the most prevailing and costly complication of MDD. Depressive symptoms impair the person's concentration, problem-solving, motivation, and energy (Masuyama, 2025). This condition occurs at the cost of the person's presence in his/her place of work and can be identified in terms of absenteeism, number of working days, and presenteeism, productivity loss at the cost of working days, as this individual attends his/her place of work. Even moderate depression can pose substantial effects upon an individual in relation to his/her working place, particularly in tasks requiring cognitive complexity and social interaction.

3.2.3.2 Social Participation

Sociational functionality is also severely affected due to MDD as patients tend to withdraw from family and societal relationships. Lack of responsiveness to emotions, feelings of worthlessness, and sensitivity to rejection are considered variables that usually lead to difficulties in relationships (Liao et al., 2025). Sociational isolation is not only considered to be the major site of dysfunction, but also the symptoms of depression tend to worsen as there's a cycle between sociational dysfunction and the intensity of episodes of depression. Sociational dysfunction also leads to limitations within lay support, which plays an essential role during the recovery process.

3.2.4 Activities of Daily Living

There is also major depression when one is not capable of undertaking basic and instrumental activities for daily living. One will not be in a position to handle personal hygiene, housekeeping, managing funds, and following health instructions (Vita et al., 2025). Psychomotor slowing, tiredness, and executive function disturbance decrease autonomy and independence-especially in older persons or those with comorbid physical disease. These functional impairments are great contributors to a compromised quality of life and additive burdens among caregivers.

3.2.5 Disability Measurement in Major Depressive Disorder

The ability to gauge or determine disability is critical when it comes to decision-making, as it is also important when it comes to health issues. Some scales that have been used to determine functional impairment triggered by MDD include some instruments that have proven their worth. The World Health Organization Disability Assessment Schedule 2.0, or WHODAS 2.0, is a generic instrument for disability assessment that is grounded in the International Classification of Functioning, Disability, and Health and encompasses disability in six domains: cognition, mobility, self-care, getting along, life activities, and participation (Federici et al., 2017). WHODAS measures are strongly linked with disease severity and future outcome in depression. Sheehan Disability Scale (SDS) is a brief, patient-rated questionnaire that evaluates the impairment level influenced by a condition in work/school, social, and family functioning (Correll et al., 2025). Its advantages lie in its ease of use, making it an important drug-evaluation tool as well as an effective investigational instrument to evaluate functional change as a secondary outcome, apart from symptom reduction. The Work and Social Adjustment Scale (WSAS) is specifically concerned with the effects of mental health difficulties in relation to work, household management, social leisure activities, private leisure activities, and intimate relationships (Hovmand et al., 2024). It is particularly helpful in the assessment of psychological or social therapies and recovery in relation to functionality, as it is a measure of role performance rather than symptoms. These measures collectively represent an acceptance that assessment of disability should be a holistic aspect of mood disorder depression treatment.

3.3 Burden and Domains of Disability in Major Depressive Disorder

Major Depressive Disorder (MDD) is rapidly being considered to be not only a mood disorder but also a disabling condition that affects individuals pervasively and persistently (Marx et al., 2023). The disability that arises due to this disease affects various activities and areas, and these include occupational functioning, interpersonal functioning, intellectual functioning, and age-related societal functioning. It is also to be noted that this disability can persist for a long time, even after partial remission of depressive symptoms. One domain which has been found to be most drastically influenced by MDD patients is occupational function. The disability caused by depression can be best explained by the presence of absence, poor work abilities, poor decision-making skills, and poor motivation. But it has been observed that the problem of presenteeism, which refers to the reduced work efficiency of the worker who can actually come to work, adds up to a major problem than the problem of absenteeism (Devita et al., 2022). MDD patients have been found to have problems related to concentration, initiation, and maintenance, resulting in decreased work efficiency and increased errors made by the worker. This has a major impact on the occupation-related progress and stability. Social disability is another cardinal component of MDD and significantly compromises quality of life. Depressed persons often indulge in social withdrawal, reduced engagement in social and interpersonal interactions, and decreased ability for mutual emotional expression (Vivekananthan & Ponnusamy, 2025). Negative thinking style, sensitivity to rejection, and ideas of worthlessness compromise the ability to form and retain social and interpersonal relationships. Social isolation interrupts the cycle of depression. This disrupted interpersonal function has important repercussions related to both the family role and the caring role. Cognitive impairment is an essential but often neglected aspect of MDD-related disability. Issues of attention, working memory, speed of processing, and executive function deficits can also be very common and can be present even after the remission of depression. The implications of cognitive deficits can be varied and range from issues of everyday tasks and learning and work performance, implying deficits of function independently, and further contribute to deficits of motivation due to action monitoring and planning deficits. Caught within the treatment-resistant and transdiagnostic concept of MDD-related disability is the concept of cognitive impairment and MDD.

A lot of difficulties arise as one grows, such as in adolescents, the disability caused by MDD can affect learning, interpersonal relationships, and the development of identity. School absences, failure in studies, and problems with concentration can influence academic and working life in the long run. Social isolation at such a critical life stage can lead to problems in learning interpersonal relationships, which in turn can lead to more episodes of mental illness. In terms of individuals in the working age group, the disability caused by MDD can severely affect working, parenting, and contributing to society (Del Casale et al., 2021). Loss of function leads to functional impairment in the workforce, interpersonal relationships, and problems in

Table 3.2 Impact of comorbid conditions on disability in MDD

Comorbidity	Mechanism of increased disability	Functional impact
Anxiety disorders	Heightened avoidance, hypervigilance	Severe social and occupational impairment
Substance use disorders	Impaired cognition, motivation, self-regulation	Poor treatment adherence, social instability
Chronic medical illness	Bidirectional symptom exacerbation	Reduced physical and mental functioning
Cognitive decline	Reduced compensatory capacity	Persistent disability despite mood improvement

contributing to society. This, in turn, can lead to disadvantage in society, overutilization of healthcare services, and more episodes of depression. In the geriatric segment, the disability caused by MDD can occur in association with the presence of illness, mental impairment, and social isolation. In elderly patients, MDD can lead to problems in performing day-to-day activities, loss of independence, and dependence on others. Institutionalization and mortality can become a reality in the geriatric segment with functional disability due to MDD. The impairment suffered by those with MDD can be further aggravated by the presence of psychiatric and medical comorbidities. The co-existence of other forms of anxiety disorders can lead to increased levels of avoidant behavior and impairments in function (Table 3.2) (Ma & Jia, 2024). The presence of substance use disorders further obscures executive control and motivational levels and can impair social and occupational function levels. Chronic medical illnesses such as diabetes, cardiovascular diseases, and chronic pain conditions can also have mutually interactive and adverse effects on depressive as well as functional levels.

3.4 Neurobiological and Psychosocial Underpinnings of Disability in Major Depressive Disorder

The Disability in Major Depressive Disorder arises from a synergy that has a profoundly complex interplay between neurobiological dysfunction and psychosocial circumstances (Berk et al., 2023). Though symptoms such as sadness and anhedonia are of clinical significance in depression, today there is a growing recognition of the importance of functional problems in such areas as cognition and motivation, social relations, and work function being linked to impaired performance of specific neural circuits in view of extrinsic stressors. It is necessary to understand these to appropriately design mental healthcare strategies that aim toward recovery in functioning, in addition to symptoms, as shown in Fig. 3.2.

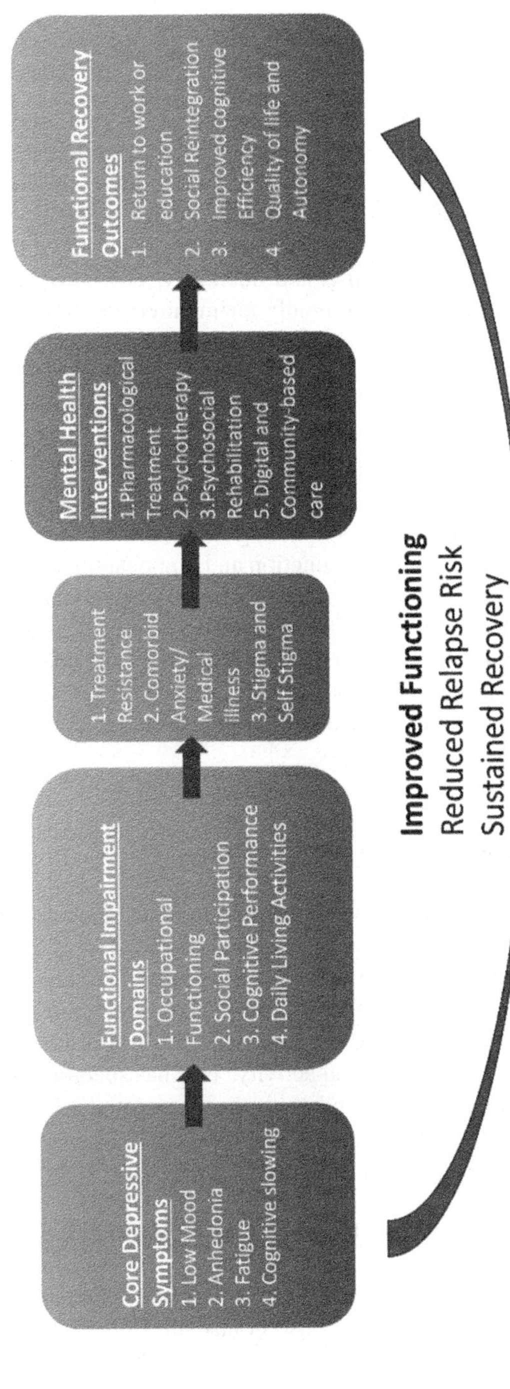

Fig. 3.2 From symptoms to functional recovery in major depressive disorder

3.4.1 Neural Circuits Implicated in Functional Impairment

3.4.1.1 Prefrontal Cortex Dysfunction

The Prefrontal Cortex (PFC), especially the dorsolateral and ventromedial parts, has a significant role in executive processes, decision-making, control of emotion, and goal-driven behavior (Fine & Hayden, 2022). In MDD, there are consistencies in alterations in the PFC concerning the whole grey matter volume and PFC interconnectivity with subcortical areas. These aspects have contributed to a deficiency in attention, working memory, planning, and flexibility. These are direct results of occupational and social disability. Additionally, an impaired top-down system control from the PFC has led to an issue concerning the regulation of adverse emotions.

3.4.1.2 Limbic System Hyperreactivity

The limbic system, which includes regions like amygdala, hippocampus, and anterior cingulate cortex, is especially crucial for handling emotions and responding to stress. In a patient with MDD, amygdala hyperfunction and hippocampus dysfunction have been implicated with heightened emotional responsivity, negativity, and difficulties with adapting to stress (Hossein et al., 2023). Such abnormalities might impact interpersonal relationships and enhance other avoidant behaviors and additional everyday functions. The hippocampus dysfunction, which is usually linked with higher levels of glucocorticoid exposure and stress, is also responsible for issues with memory, thereby adding to disability.

3.4.1.3 Neurotransmission and Reward Circuitry

Problems with the reward circuitry system within the brain, specifically the ventral striatum, nucleus accumbens, or dopamine systems, account for the problems with anhedonia and motivational deficits found in MDD patients (Bekhbat et al., 2022). Blunt reward responsiveness and problems with reward-directed learning decrease the person's ability to participate and maintain efforts toward attaining a goal or activity, such as work, self-care, or social activity. This has specific effects on functional outcomes because motivation and reward are essential for productivity or participation with the intervention.

3.4.1.4 Role of Inflammation and Neuroendocrine Dysregulation

Apart from the circuits, the biological aspects of the body, like the inflammatory process and the neuroendocrine system, also contribute to the levels of disability caused by MDD. There is growing evidence suggesting the presence of low-grade inflammation in the body, as a consequence of which increased levels of pro-inflammatory cytokines are observed, and MDD is found to be intimately associated with the inflammatory process (Toenders et al., 2022). Neuroendocrine regulation, especially with respect to the Hypothalamic Pituitary Adrenal Axis, is also a defining characteristic of people with MDD. Chronic over-activation of this pathway results in chronically high levels of cortisol, which impact crucial areas of cognition and regulation in a deleterious manner (Zhu et al., 2021). Other aspects of regulation, such as physical energy levels, sleep, and stress, that are considered important in determining functional status, are made vulnerable by virtue of this aforementioned axis.

3.4.1.5 Psychosocial Contributors to Disability

Exposure to early-life stress, such as trauma, neglect, or adverse experiences, has been shown to be a risk factor for the development and severity of MDD. Early-life stress can affect brain development and increase vulnerability to disability due to difficulties with stress responsiveness and emotional regulation (Nakama et al., 2023). These individuals tend to have a stress sensitivity problem, which makes them resort to maladaptive coping mechanisms, leading to challenges with responding to treatments, which may contribute to disability.

Poverty, unemployment, low educational attainment, and unstable housing status all impact the functional course of MDD. These challenges are barriers to mental health service access, increase chronic exposure to stressful events, and provide poor opportunities for social and occupational reintegration. In resource-deficient contexts, the presence of disability further exacerbates the weak mental health system and poor social support structures. Additionally, stigma constitutes another significant psychosocial barrier to recovery in MDD (Kim & Ahn, 2025). Public stigma promotes discrimination, while internalized self-stigma affects one's esteem, efficacy, and intention to seek care. Social isolation because of public stigma, in combination with undertreatment, contributes to functional impairment even as one experiences disablement (McDaniels et al., 2023). Notably, the interrelations between neurobiological and psychosocial factors do not only occur from the neurobiological to the psychosocial aspect. On one hand, factors such as economic or socioeconomic problems can exacerbate biological vulnerabilities due to the stimulation of an inflammatory or stress response. On the other hand, biological vulnerabilities related to cognition and motivation can impair an individual's ability to cope with psychosocial problems, thereby creating an interrelation between disabilities in both aspects that influence each other. Notably, this underlines the significant role of an integrated approach in the context of mental health.

3.5 Mental Health Approaches to Reducing Disability in MDD

The reduction of disability in MDD patients requires an approach in treatment beyond the mere alleviation of depression symptoms, focusing on different dimensions of disability. Mental health treatment types that emphasize the reduction of disability comprise functional recovery, quality of life, and participation in daily occupations (Lianur et al., 2025). The different types of intervention in these approaches comprise biological, psychological, psychosocial, technological, and community interventions.

3.5.1 Pharmacological Interventions

3.5.1.1 Antidepressants and Functional Outcomes

The pharmacological approach has been a mainstay in treating MDD, and antidepressants have been shown to be effective in improving depressed symptoms. Nonetheless, a growing body of evidence points out that symptomatic remission does not necessarily herald functional recovery. Selective serotonin reuptake inhibitors, serotonin and norepinephrine reuptake inhibitors, and atypical antidepressants have been found to produce small gains in work function and social participation, especially in long-term treatment maintenance (Edinoff et al., 2021). The Sheehan Disability Scale and WHODAS 2.0 functional outcomes have shown that functional deficits persist despite symptomatic remission.

3.5.1.2 Cognitive and Motivational Recovery

Cognitive impairment in the form of attention, memory, and executive processing deficits also imparts significant disability to MDD patients. Some antidepressants, such as vortioxetine and bupropion, have shown relative efficacy in improving the patient's performance in these areas, which in turn may boost functional ability (Chokka et al., 2025). The enhancement of motivational or goal-direction deficits, however, has pivotal significance for the rehabilitation of the patient, and drug action in these areas has remained inconsistent.

3.5.1.3 Limitations in Disability Reduction

Even with the progress observed in the area of psychopharmacology, the employment of antidepressants has often proven ineffective in completely facilitating functional restoration by themselves (Zheng et al., 2025). Residual symptoms, the resistance of these symptoms to various treatments, the side effects of medication, and the delay of action are factors that impede the achievement of complete disability reduction.

3.5.2 *Psychotherapeutic Approaches*

Cognitive Behavioral Therapy (CBT) is one of the most extensively researched psychotherapeutic modalities that has been effective for persons experiencing MDD and has been shown to relieve both the depression and functions related to it. Through countering maladaptive cognition and behavior, CBT promotes improvement regarding problem-solving, work performance, and social functioning (Matthys & Schutter, 2022). It appears that CBT can bring long-term benefits to diminishing disability, especially if skill acquisition and relapse prevention are emphasized. Behavioral activation involves participation in meaningful and rewarding activities to counteract avoidance and inactivity, which are core components of disability in people with MDD. Research evidence shows that behavioral activation is highly effective in enhancing functional performance as well as social participation, even for people with severe depression. Behavioral activation is also highly effective because of its simplicity. Interpersonal Therapy (IPT) aims to improve interpersonal relationships and social role functioning, which are usually compromised in MDD patients (Bian et al., 2023). Through the resolution of role transitions, interpersonal disputes, and grief, improvement in social functioning and decreased disability can be observed. IPT has proven very useful for patients whose episodes of depression are closely tied to social stressors. Acceptance and Commitment Therapy (ACT) encourages psychological flexibility through acceptance of difficult experiences and value-based behavior. Some preliminary evidence indicates that an ACT approach may be effective in functional outcomes by assisting patients in integrating with life activities despite the presence of their symptoms (Han et al., 2023). This model is gaining popularity for chronic or treatment-resistant depression because disabilities often remain even after addressing their depressive symptoms.

3.5.3 *Psychosocial Rehabilitation*

Vocational rehabilitation restores work capacity and employment participation, which are important elements of functional recovery in MDD. Skill development, workplace accommodation, and gradual return-to-work strategies have been developed to enhance employment outcomes and decrease work-related disability among individuals with mental illness. Most supported employment programs, including the IPS model, place a premium on rapid job placement with ongoing support rather than extended pre-employment training. Although it was originally conceived to treat severe mental illnesses, it has been shown some recent promise in improving employment outcomes in persons who suffer from depression related disability (Riley et al.,

2021). Deficits in social skills are a contributing factor in both the poor interpersonal relationships and social withdrawal exhibited in MDD. Structured social skills training enhances communication, assertiveness, and the ability to preserve relationships. Family involvement is among the most important aspects when it comes to functional recovery, particularly in cultures that provide strong and supportive family environments. Psychoeducation, supportive care for caregivers, and family therapies have all shown to be associated with improved compliance, the reduction of relapses and improved functional outcomes.

3.5.4 Digital and Technology-Assisted Interventions

3.5.4.1 Telepsychiatry

Telepsychiatry has helped improve access to care, especially for the mobility-impaired population and those in rural areas. Findings from various studies have indicated that telepsychiatric treatments are equivalent to direct care in the improvement of depressive symptoms and function (Sharma et al., 2025).

3.5.4.2 Mobile Mental Health Applications

Mobile apps related to mood tracking, activity planning, and cognitive restructuring provide scalable interventions for persons living with disability due to MDD. Though limited in symptom relief, there has been some evidence of improved functionality and self-management capabilities (Diano et al., 2023).

3.5.4.3 Internet-Based CBT

Internet-based CBT has shown efficacy in reducing depressive symptoms as well as improving functional impairment in people with mild to moderate depression. It is a flexible and inexpensive treatment modality and an excellent addition to a stepped-care model of treatment (Lin et al., 2023).

3.5.4.4 Digital Monitoring of Functioning

Wearable technology as well as digital phenotyping techniques allow real-time tracking of activity, sleep, as well as social interaction. The technical tools provide promise in earlier detection of functional decline, as well as tailoring strategies of intervention, though issues of ethics as well as privacy persist (Jung et al., 2025).

3.5.5 Community and Public Mental Health Approaches

Task-shifting and stepped care strategies where nonspecialist health providers deliver evidence-based interventions have proven successful in preventing disability within low- and middle-income countries. Stepped care models ensure that resources are utilized efficiently by matching the level of intervention needed with the level of functionality (Bolton et al., 2023). Community mental health services stress the importance of continuity of care, inclusion, and rehabilitation. Community mental health services are imperative in caring for long-term disability through independent living and community participation (Harvey et al., 2023). Disability-Inclusive Mental Health public policies, incorporating the view of people with disabilities, ensure ease of treatment, accommodating people at their workplaces, and social security for people with MDD. Making mental health care conform to the notion of people with disabilities, like the WHO's ICF, is crucial in enabling effective functional recovery (Jain et al., 2025).

3.6 Integrated and Recovery-Oriented Care Models in Major Depressive Disorder

Integrated and recovery-oriented approaches to care mark the shift in how Major Depressive Disorder is managed today, as they have transitioned from only concentrating on the symptom relief associated with MDD to one focused on the recovery of functionality (Sampogna et al., 2022). Integrated care is also reflective of the recognition that MDD is associated with disability that is not one- or two- dimensional; it is multifaceted in that it involves occupational function, interpersonal relationships, cognitive function, and quality-of-life considerations.

3.6.1 Collaborative Care Models

Collaborative Care is one of the oldest and most extensively investigated integrated strategies used in the management of MDD and has been proven to be effective in the reduction of both the depressive symptoms and the functional disability caused by the disorder (Reist et al., 2022). The main features of this approach include a systematic collaboration between the primary care professionals, specialists, and the patient's care manager. Some of the features of this approach include a structured treatment approach, continuous evaluation of the outcomes, and the appropriate adjustments of the stepped care approach according to the responses of the patients. The main concept of this approach is the integration of mental health care within the overall healthcare systems, an approach which promotes easier accessibility and compliance, especially in cases of patients whose disability due to MDD makes it difficult for the patients to access the required health care (Zhong et al., 2022).

3.6.2 Personalized Treatment Planning

Personalized treatment plans are an integral part of recovery-oriented treatment. Unlike standardized treatment protocols based on symptoms of illness, personalized treatment plans take into consideration the unique manifestations of illness, functional deficits, comorbid problems, psychosocial issues, and patient preferences (Menear et al., 2022). Functional assessment instruments like work performance measures, social functioning measures, and measures of cognitive capacity are gradually receiving equal importance in treatment planning alongside instruments that measure symptoms of illness. Such an individualized treatment plan will help in addressing areas of disability which are of utmost significance to the patient. Such areas include resuming the patient's occupation or resuming social roles or enhancing the activities of day-to-day living through treatment. Such an individualized treatment plan assists in improving patient engagement and empowerment in treatment plans, which is of immense significance in the treatment of a patient.

3.6.3 Emphasis on Functional Goals

Recovery-oriented models of treatment emphasize functional objectives more than before, inasmuch as functional recovery may not be the natural byproduct of symptomatic recovery. Functional objectives may encompass returning to work or schooling, enhancing relational functions, or developing everyday self-sufficiency. The implementation of behavioural activation techniques, rehabilitation therapies, and cognitive repair therapies can help tackle these goals in combination (Ciharova et al., 2021). In this fashion, instead of focusing purely on recovery milestones of symptoms, treatment gains can be measured and quantified in terms of how these objectives are achieved. In this fashion, there can also be a new standard of measuring treatment success through recovery-oriented models.

3.6.4 Quality of Life as a Central Outcome

The aspect of quality of life has become important in caring for MDD-related disabilities. Recovery-focused care acknowledges the fact that addressing symptoms does not automatically mean there will be any changes in how patients feel about life. Well-being, having life purpose, autonomy, and relatedness are becoming important elements in the journey to recovery. Strategies like psychosocial interventions, social support strategies, as well as life-style interventions are becoming important in changing the quality of life in patients who still exhibit some symptoms related to depression.

3.6.5 *Social Reintegration*

Social re-integration represents another important component of a model of care with a focus on recovery, especially in consideration of the extensive degree of social withdrawal and role disruption occurring with MDD (Meyer et al., 2024). Stigmatizing social identity and extensive social disengagement from work or social life may contribute to sustained disability even in patients showing significant improvement in their condition. Integrated models of care offer appropriate solutions for those problems by using supported employment interventions and family interventions along with peer and social rehabilitation interventions.

3.6.6 *Role of Multidisciplinary Teams*

Multidisciplinary teams are a crucial part of the effective implementation of care. Multidisciplinary teams should include psychiatrists, psychologists, psychiatric social workers, occupational therapists, nurses, and of course, peer support specialists. Each one of these has something uniquely different to offer in managing the complexities of the biological, psychological, or social components of the disability associated with MDD. Teamwork is valuable and encourages comprehensive approaches to intervention. There is also increased scope within multidisciplinary working when treating comorbidities, and also aspects of the social model of health that are beyond the remit of normal care.

3.7 Challenges, Gaps, and Future Directions

Although there has been progress in the treatment of Major Depressive Disorder (MDD), there are many challenges in the treatment of the induced disability caused by such an illness. One of the most serious issues is treatment-resistant depression, in which patients remain functionally disabled despite reaching partial or full symptomatic remission of the illness. The co-existence of persistent disability in such patients indicates the lack of effectiveness of treatments that are primarily focusing on symptoms and thus signal the importance of addressing cognitive, occupational, and social functions explicitly in clinical practice. In this context, it can also be noted that there are serious research gaps in measuring functional outcomes in clinical trials in which there is a lack of utilization of standardized tools for measuring disabilities, which may efficiently portray the real-life functions of patients in everyday life. Disparities in terms of cultural and socioeconomic parameters are also responsible for complicating this situation in low and middle-income countries in which mental health stigma, lack of availability of mental health services, and lack of equality in mental health fare worse due to which patients remain functionally disabled for many years. To overcome such research gaps in clinical trials for treating those associated disabilities in individuals with MDD, there should be a considerable emphasis in research and clinical trials related to longitudinal studies in which the trajectory of functional impairment caused due to this illness may be defined properly. In this context, for treating disabilities in individuals associated with this health problem, there should also be considerable clinical trials in which the treatment should not remain focused only on symptoms. In this context, selected mental health strategies should not remain just focused on treating symptoms of mental health issues but should remain focused more on overall functions of individuals in everyday life. In this context, intervention in mental health treatment strategies may remain focused in which associated mental health treatment strategies are incorporated within primary health systems for treating associated disabilities in individuals associated with this health problem.

References

Almakrob, A. Y., & Alduais, A. (2025). Depression and Anxiety in the Saudi Population: Epidemiological Profiles from Health Surveys and Mental Health Services. *INQUIRY: The Journal of Health Care Organization, Provision, and Financing, 62*, 00469580251382027.

Bekhbat, M., Li, Z., Mehta, N. D., Treadway, M. T., Lucido, M. J., Woolwine, B. J., Felger, J. C., et al. (2022). Functional connectivity in reward circuitry and symptoms of anhedonia as therapeutic targets in depression with high inflammation: Evidence from a dopamine challenge study. *Molecular Psychiatry, 27*(10), 4113–4121.

Berk, M., Köhler-Forsberg, O., Turner, M., Penninx, B. W., Wrobel, A., Firth, J., Marx, W., et al. (2023). Comorbidity between major depressive disorder and physical diseases: A comprehensive review of epidemiology, mechanisms and management. *World Psychiatry, 22*(3), 366–387.

Bian, C., Zhao, W. W., Yan, S. R., Chen, S. Y., Cheng, Y., & Zhang, Y. H. (2023). Effect of interpersonal psychotherapy on social functioning, overall functioning and negative emotions for depression: A meta-analysis. *Journal of Affective Disorders, 320*, 230–240.

Bolton, P., West, J., Whitney, C., Jordans, M. J., Bass, J., Thornicroft, G., Raviola, G., et al. (2023). Expanding mental health services in low-and middle-income countries: A task-shifting framework for delivery of comprehensive, collaborative, and community-based care. *Cambridge Prisms: Global Mental Health, 10*, Article e16.

Chokka, P., Hammer-Helmich, L., Schmidt, S. N., Hubert, M., Reines, E. H., & Grande, I. (2025). Functional improvement as a treatment goal in major depressive disorder: A narrative review of the evidence for vortioxetine. *Current Medical Research and Opinion, 41*(5), 855–866.

Ciharova, M., Furukawa, T. A., Efthimiou, O., Karyotaki, E., Miguel, C., Noma, H., Cuijpers, P., et al. (2021). Cognitive restructuring, behavioral activation and cognitive-behavioral therapy in the treatment of adult depression: A network meta-analysis. *Journal of Consulting and Clinical Psychology, 89*(6), 563.

Correll, C. U., Cortese, S., Solmi, M., Boldrini, T., Demyttenaere, K., Domschke, K., McIntyre, R. S., et al. (2025). Beyond symptom improvement: Transdiagnostic and disorder-specific ways to assess functional and quality of life outcomes across mental disorders in adults. *World Psychiatry, 24*(3), 296–318.

Del Casale, A., Zocchi, C., Kotzalidis, G. D., Fiaschè, F., & Girardi, P. (2021). Prevention of depression in children, adolescents, and young adults: The role of teachers and parents. *Psychiatry International, 2*(3), 353–364.

Derblom, K., Dahlberg, K., Gabrielsson, S., Lindgren, B. M., & Molin, J. (2025). Key Aspects of Recovery-Oriented Practice in Caring for People With Mental Ill-Health in General Emergency Departments: A Modified Delphi Study. *Journal of Clinical Nursing, 34*(2), 565–579.

Devita, M., De Salvo, R., Ravelli, A., De Rui, M., Coin, A., Sergi, G., & Mapelli, D. (2022). Recognizing depression in the elderly: practical guidance and challenges for clinical management. *Neuropsychiatric Disease and Treatment*, 2867–2880.

Diano, F., Sica, L. S., & Ponticorvo, M. (2023). Empower psychotherapy with mHealth apps: The design of "Safer", an emotion regulation application. *Information, 14*(6), 308.

Early-life stress can affect brain development and increase vulnerability to disability due to difficulties with stress responsiveness and emotional regulation.

Edinoff, A. N., Akuly, H. A., Hanna, T. A., Ochoa, C. O., Patti, S. J., Ghaffar, Y. A., Kaye, A. M., et al. (2021). Selective serotonin reuptake inhibitors and adverse effects: A narrative review. *Neurology International, 13*(3), 387–401.

Federici, S., Bracalenti, M., Meloni, F., & Luciano, J. V. (2017). World Health Organization disability assessment schedule 2.0: An international systematic review. *Disability and rehabilitation, 39*(23), 2347–2380.

Fine, J. M., & Hayden, B. Y. (2022). The whole prefrontal cortex is premotor cortex. *Philosophical Transactions of the Royal Society B: Biological Sciences, 377*(1844).

Hagan, M., Hernandez, B., & Batchelder, A. W. (2025). Self-Worth in Mental Health: At the Intersection of Mattering and Significance. In *The Routledge International Handbook of Human Significance and Mattering* (pp. 275–288). Routledge.

Han, A., Wilroy, J. D., & Yuen, H. K. (2023). Effects of acceptance and commitment therapy on depressive symptoms, anxiety, pain intensity, quality of life, acceptance, and functional impairment in individuals with neurological disorders: A systematic review and meta-analysis. *Clinical Psychologist, 27*(2), 210–231.

Harvey, C., Zirnsak, T. M., Brasier, C., Ennals, P., Fletcher, J., Hamilton, B., Brophy, L., et al. (2023). Community-based models of care facilitating the recovery of people living with persistent and complex mental health needs: A systematic review and narrative synthesis. *Frontiers in Psychiatry, 14*, 1259944.

Hossein, S., Cooper, J. A., DeVries, B. A., Nuutinen, M. R., Hahn, E. C., Kragel, P. A., & Treadway, M. T. (2023). Effects of acute stress and depression on functional connectivity between prefrontal cortex and the amygdala. *Molecular Psychiatry, 28*(11), 4602–4612.

Hovmand, O. R., Reinholt, N., Bryde Christensen, A., Bach, B., Eskildsen, A., Arendt, M., Arnfred, S. M., et al. (2024). Utility of the Work and Social Adjustment Scale (WSAS) in predicting long-term sick-leave in Danish patients with emotional disorders. *Nordic Journal of Psychiatry, 78*(1), 14–21.

Jain, P., Jain, B., Doshi, R., Jain, U., Claypool, H., Aboulafia, A., & Swenor, B. K. (2025). *Digital health: An opportunity to advance health equity for people with disabilities.*

Jung, H. W., Kim, D. Y., Lee, I., Kim, O., Lee, S., Lee, S., Lee, J. J., et al. (2025). Key features of digital phenotyping for monitoring mental disorders: Systematic review. *Journal of Medical Internet Research, 27*, Article e77331.

Kim, Y. V., & Ahn, H. (2025). Protecting the Mental Health of Esports Players: A Qualitative Case Study on Their Stress, Coping Strategies, and Social Support Systems. *International Journal of Mental Health Promotion, 27*(9).

Lianur, L., Haryani, A., & Putri, M. I. (2025). The Effect of Occupational Therapy on Increasing Functional Independence in Patients with Mental Disorders. *International Journal on ObGyn and Health Sciences, 3*(2), 38–48.

Liao, Z. L., Pu, X. L., Zheng, Z. Y., & Luo, J. (2025). Social function scores and influencing factors in patients with residual depressive symptoms. *World Journal of Psychiatry, 15*(1), 98630.

Lin, Z., Cheng, L., Han, X., Wang, H., Liao, Y., Guo, L., McIntyre, R. S., et al. (2023). The effect of internet-based cognitive behavioral therapy on major depressive disorder: Randomized controlled trial. *Journal of Medical Internet Research, 25*, Article e42786.

Ma, S., & Jia, N. (2024). The Symptom Structure and Causal Relationships of Comorbid Anxiety and Depression Among Chinese Primary and Middle School Teachers: A Network Analysis. *Psychology Research and Behavior Management*, 3731–3747.

Marx, W., Penninx, B. W., Solmi, M., Furukawa, T. A., Firth, J., Carvalho, A. F., & Berk, M. (2023). Major depressive disorder. *Nature Reviews Disease Primers, 9*(1), 44.

Masuyama, A. (2025). Evolutionary explanations of depression and cognitive control dysfunction: A literature review. *Behavioural Brain Research*, 116001.

Matthys, W., & Schutter, D. J. (2022). Improving our understanding of impaired social problem-solving in children and adolescents with conduct problems: Implications for cognitive behavioral therapy. *Clinical Child and Family Psychology Review, 25*(3), 552–572.

McDaniels, B., Pontone, G. M., Mathur, S., & Subramanian, I. (2023). Staying hidden: The burden of stigma in PD. *Parkinsonism & Related Disorders, 116*, Article 105838.

Menear, M., Girard, A., Dugas, M., Gervais, M., Gilbert, M., & Gagnon, M. P. (2022). Personalized care planning and shared decision making in collaborative care programs for depression and anxiety disorders: A systematic review. *PLoS ONE, 17*(6), Article e0268649.

Meyer, A. F., Casteleijn, D., & Silaule, O. (2024). The therapeutic impact of occupational therapy groups on the activity participation of persons with major depressive disorder in an acute mental health setting. *South African Journal of Occupational Therapy, 54*(2), 18–26.

Nakama, N., Usui, N., Doi, M., & Shimada, S. (2023). Early life stress impairs brain and mental development during childhood increasing the risk of developing psychiatric disorders. *Progress in Neuro-Psychopharmacology and Biological Psychiatry, 126*, Article 110783.

Rao, S., Dimitropoulos, G., Milaney, K., Eurich, D. T., Patten, S. B., & HEARTS Study Core-searchers. (2025). Mental health-related disabilities, healthcare utilization, and access in young adults with anxiety and depression: a scoping review protocol. *Social Work in Mental Health*, 1–26.

Reist, C., Petiwala, I., Latimer, J., Raffaelli, S. B., Chiang, M., Eisenberg, D., & Campbell, S. (2022). Collaborative mental health care: A narrative review. *Medicine, 101*(52), Article e32554.

Riley, J., Drake, R. E., Frey, W., Goldman, H. H., Becker, D. R., Salkever, D., Karakus, M., et al. (2021). Helping people denied disability benefits for a mental health impairment: The supported employment demonstration. *Psychiatric Services, 72*(12), 1434–1440.

Sağbaş, S., Korkmaz, ŞA., & Öyekçin, D. G. (2025). The Effect of Brief Group Psychoeducation on Cognitive Distortions, Automatic Thoughts and Functioning in Major Depressive Disorder: A Randomized Controlled Trial. *Clinical Psychology & Psychotherapy, 32*(5), Article e70165.

Sampogna, G., Di Vincenzo, M., Giallonardo, V., Luciano, M., & Fiorillo, A. (2022). Recovery-oriented treatments in Major depressive disorder. *Recovery and major mental disorders* (pp. 245–254). Springer International Publishing.

Sharma, M., Sharma, V., & Peeples, D. (2025). Telepsychiatry, access, and equity: Accelerating mental health care for rural and low-income youth. *Frontiers in Public Health, 13*, 1698682.

Subedi, T. N. (2024). An Integrative Approach to Sociology of Disability: A Theoretical Recommendation. *SMC Journal of Sociology, 1*(1), 1–23.

Toenders, Y. J., Laskaris, L., Davey, C. G., Berk, M., Milaneschi, Y., Lamers, F., Schmaal, L., et al. (2022). Inflammation and depression in young people: A systematic review and proposed inflammatory pathways. *Molecular Psychiatry, 27*(1), 315–327.

Vita, G., Magro, V. M., Sorbino, A., Ljoka, C., Manocchio, N., & Foti, C. (2025). Opportunities Offered by Telemedicine in the Care of Patients Affected by Fractures and Critical Issues: A Narrative Review. *Journal of Clinical Medicine, 14*(20), 7135.

Vivekananthan, K., & Ponnusamy, R. (2025). Mental health of the empty nest elderly. In *Handbook of Aging, Health and Public Policy: Perspectives from Asia* (pp. 1277–1298). Singapore: Springer Nature Singapore.

Yan, G., Zhang, Y., Wang, S., Yan, Y., Liu, M., Tian, M., & Tian, W. (2024). Global, regional, and national temporal trend in burden of major depressive disorder from 1990 to 2019: An analysis of the global burden of disease study. *Psychiatry Research, 337*, Article 115958.

Zhang, W., & Wang, A. P. (2025). Functional ability of older adults based on the World Health Organization framework of healthy ageing: A scoping review. *Journal of Public Health, 33*(7), 1513–1531.

Zheng, Y., Ma, Y., Pan, Y., Ali, T., Zheng, C., Miu, K. K., Tan, Z., et al. (2025). Autophagy modulation by antidepressants: Mechanisms and implications. *Neurochemical Research, 50*(5), 285.

Zhong, B. L., Xu, Y. M., & Li, Y. (2022). Prevalence and unmet need for mental healthcare of major depressive disorder in community-dwelling Chinese people living with vision disability. *Frontiers in Public Health, 10*, Article 900425.

Zhu, Y., Wu, X., Zhou, R., Sie, O., Niu, Z., Wang, F., & Fang, Y. (2021). Hypothalamic-pituitary-end-organ axes: Hormone function in female patients with major depressive disorder. *Neuroscience Bulletin, 37*(8), 1176–1187.

Chapter 4
Systems-Based Integration of Disability Care in Depressive Disorders

Niharika Singh, Hareni R, and Rajamerlin E

Abstract Depressive disorders are a leading cause of disability worldwide, contributing substantially to functional impairment, reduced quality of life, and socioeconomic burden across populations. Although effective pharmacological and psychological treatments are available, disability care for depressive disorders remains inadequately addressed due to fragmented mental health systems and limited coordination across services. Such fragmentation often results in discontinuity of care, delayed rehabilitation, and poor long-term functional recovery. This chapter examines the importance of systems-based integration in addressing disability associated with depressive disorders, emphasizing coordinated healthcare delivery across clinical, rehabilitative, community, and social support domains. Integrated care models, including primary–secondary–tertiary healthcare linkages, collaborative care, and stepped-care frameworks, are discussed as effective strategies for improving access, continuity, and outcomes in depressive disability care. The role of multidisciplinary mental health teams in promoting interprofessional collaboration and shared decision-making is also highlighted. The chapter further explores the contribution of community-based and institutional support systems, including rehabilitation services, non-governmental organizations, and social welfare agencies, in facilitating recovery and social reintegration. Emerging digital and technological interventions, such as telepsychiatry and electronic health records, are examined for their potential to enhance care continuity and system integration. Barriers to effective integration, including structural constraints, workforce shortages, stigma, and policy limitations, are critically analyzed. Finally, strategies for strengthening integrated disability care systems through policy reform, capacity building, and service redesign are proposed, with lessons drawn from national and international models relevant to low- and middle-income settings.

N. Singh (✉) · H. R · R. E
Department of Biotechnology, VSB Engineering College, Karur, India
e-mail: nihubiotech@gmail.com

A. Kumar (ed.), *Integrating Disability Care in Major Depressive Disorder Through Neurological and Mental Health Perspectives for Empowerment*,
SpringerBriefs in Modern Perspectives on Disability Research,
https://doi.org/10.1007/978-981-95-8955-5_4

Keywords Depressive disorders · Disability care · Systems-based integration · Integrated mental health services · Community mental health · Rehabilitation · Telepsychiatry · Health systems strengthening

4.1 Introduction

Depressive disorders are among the most prevalent mental health conditions worldwide and constitute a leading cause of disability across age groups and socioeconomic contexts. Beyond emotional distress, depression significantly impairs cognitive functioning, interpersonal relationships, educational attainment, and work productivity, resulting in long-term functional limitations and reduced quality of life (Vos et al., 2020; Whiteford et al., 2013; World Health Organization, 2013, 2017). The disability associated with depressive disorders often persists even after partial symptomatic improvement, highlighting the need for care models that address both clinical and functional recovery. Depression-related disability is inherently multidimensional, shaped by biological vulnerability, psychological factors, social disadvantage, and environmental barriers. Traditional mental health services have largely emphasized symptom reduction through pharmacological and psychotherapeutic interventions, with comparatively less attention to rehabilitation, social participation, and long-term support. Such narrowly focused approaches may be insufficient for individuals experiencing chronic or recurrent depression, who frequently require coordinated medical, psychosocial, and social welfare interventions (Patel et al., 2018, Prince et al., 2007, World Health Organization and World Bank, 2011, Bhugra et al., 2013). A systems-based approach to disability care offers a structured framework to address these complexities by integrating services across different levels of care and sectors. Systems-based integration emphasizes continuity, coordination, and person-centeredness, ensuring that individuals with depressive disorders receive timely and appropriate support across the continuum of prevention, treatment, rehabilitation, and reintegration (Funk et al., 2008; Institute of Medicine, 2006; Murray & Lopez, 1996; Wagner et al., 2001). By strengthening linkages between healthcare providers, rehabilitation services, and community resources, integrated systems aim to reduce service gaps, prevent relapse, and improve long-term functional outcomes.

4.1.1 Rationale for Systems-Based Approaches in Depressive Disability Care

Systems-based approaches are grounded in the recognition that depressive disability cannot be effectively managed through isolated or episodic interventions. Depression frequently coexists with physical illnesses, unemployment, social exclusion, and stigma, all of which contribute to the severity and persistence of disability (Harvey et al., 2017; Lund et al., 2010). Integrated systems facilitate coordinated assessment, shared care planning, and longitudinal follow-up, enabling interventions to be tailored to changing clinical and functional needs over time. Such approaches also promote recovery-oriented care by shifting the focus from symptom remission alone to the restoration of autonomy, social participation, and productive functioning. Evidence from collaborative and stepped-care models demonstrates that coordinated systems can improve access to care, enhance treatment adherence, and optimize resource utilization, particularly in resource-constrained settings (Archer et al., 2012; Bower & Gilbody, 2005; Bodenheimer et al., 2002; Gilbody et al., 2006; Unitizer et al., 2002).

4.1.2 Limitations of Fragmented Mental Health Services

Fragmented mental health service delivery remains a major barrier to effective disability care in depressive disorders. Poor coordination between primary care, specialist services, rehabilitation programs, and social welfare systems often results in discontinuity of care, inconsistent treatment strategies, and inadequate monitoring of functional outcomes (Saraceno et al., 2007; Saxena et al., 2007). Individuals may experience delays in referral, repeated assessments, or premature discontinuation of care, increasing the risk of relapse and chronic disability. Moreover, fragmented systems frequently neglect psychosocial rehabilitation, vocational training, and community-based support, which are critical for sustained recovery and social reintegration. These gaps disproportionately affect vulnerable populations and contribute to widening health inequities (Corrigan et al., 2012; Dixon, 2000). Addressing these limitations requires a deliberate transition toward integrated, systems-oriented models that align clinical care with functional, social, and policy-level objectives in depressive disability care.

4.2 Concept of Integrated Disability Care Systems

Integrated disability care systems have emerged as a central framework for addressing the complex and long-term needs of individuals living with depressive disorders. Unlike conventional models that operate through isolated services, integrated systems emphasize coordinated action across healthcare, rehabilitation, and social support sectors. Table 4.1 describes such systems that aim to deliver seamless, continuous, and person-centered care that addresses both the clinical manifestations of depression and the associated functional disabilities (Patel et al., 2018, World Health Organization and World Bank, 2011). Depressive disorders frequently require long-term management due to their recurrent nature and the persistence of disability even after symptom improvement. Integrated care systems recognize this chronicity and are designed to support individuals across the full continuum of care, from early

Table 4.1 Key components of integrated disability care systems

Component	Description	Examples/Implementation strategies
Clinical Services	Core medical and psychiatric interventions for depressive disorders	Diagnosis, pharmacotherapy, psychotherapy, rehabilitation
Community Support	Engagement of local communities for ongoing psychosocial support	Peer groups, family counseling, home visits
Institutional Support	Specialized facilities providing structured care	Rehabilitation centers, day-care programs, vocational training
Workforce	Multidisciplinary team supporting patient-centered care	Psychiatrists, psychologists, social workers, rehabilitation specialists
Digital & Technological Support	Use of telepsychiatry, mobile health, and EHRs for coordination	Teleconsultation, symptom monitoring apps, integrated EHR systems

identification and treatment to rehabilitation, community reintegration, and relapse prevention. By reducing fragmentation and enhancing communication among service providers, integrated systems improve efficiency, quality of care, and functional outcomes (Institute of Medicine, 2006; Wagner et al., 2001).

4.2.1 Definition and Scope of Integrated Care

Integrated disability care can be defined as the systematic coordination of medical, psychological, rehabilitative, and social services to address the multidimensional needs of individuals with depression-related disability. The scope of integrated care extends beyond clinical treatment to include psychosocial rehabilitation, vocational support, housing assistance, and social protection measures that facilitate participation and independence (Rapp & Goscha, 2011; World Health Organization, 2010).

The World Health Organization and other global health bodies emphasize integrated care as a cornerstone of effective mental health systems, particularly in low- and middle-income countries where resources are limited and service gaps are pronounced (Funk et al., 2008; World Health Organization, 2017). Figure 4.1 explains about the integrated systems operated at multiple levels individual, service, organizational, and policy ensuring alignment between care delivery and broader health and social objectives (Institute of Medicine, 2006; Marmot et al., 2008; Patel et al., 2015).

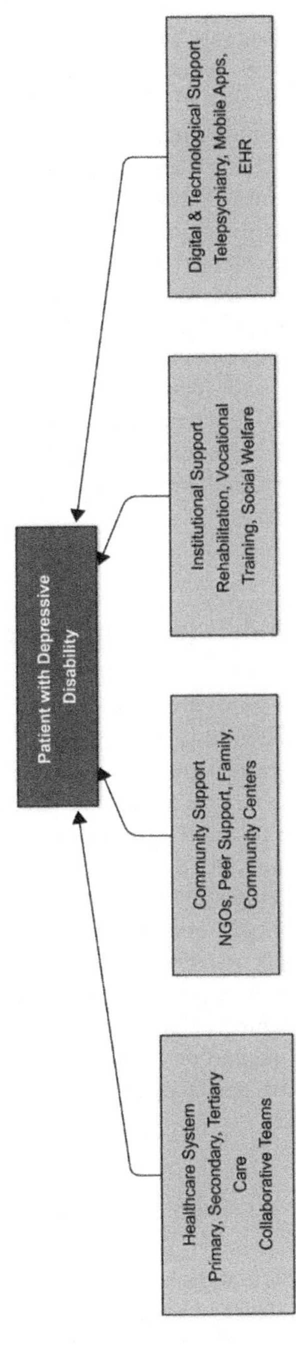

Fig. 4.1 System-based integration of disability care for depressive disorders

4.2.2 Multidisciplinary and Intersectoral Coordination

A defining feature of integrated disability care systems is multidisciplinary collaboration. Effective integration requires the coordinated involvement of psychiatrists, psychologists, primary care physicians, social workers, rehabilitation specialists, and community health workers, each contributing distinct expertise to a shared care plan (Drake et al., 2004; Patel et al., 2015; Rapp & Goscha, 2011). Regular communication, shared documentation, and collaborative decision-making are essential to avoid duplication and ensure consistency of care. In addition to multidisciplinary teamwork within the health sector, intersectoral coordination plays a critical role in addressing the social determinants of depressive disability. Collaboration with education systems, employment services, housing agencies, non-governmental organizations, and social welfare departments enables integrated systems to address barriers such as unemployment, poverty, and social exclusion that exacerbate disability (Harvey et al., 2017; Lund et al., 2010; Organisation for Economic Co-operation & Development, 2015). Through such coordinated efforts, integrated disability care systems support holistic recovery and promote long-term social inclusion for individuals with depressive disorders.

4.3 Healthcare System Models for Depressive Disability

Healthcare system models for depressive disability focus on organizing services in a manner that ensures continuity, accessibility, and effectiveness across different levels of care. Given the chronic and recurrent nature of depressive disorders and the persistence of associated disability, these models emphasize structured pathways that link prevention, treatment, rehabilitation, and long-term follow-up within an integrated system (Funk et al., 2008; Patel et al., 2018; Wagner et al., 2001). Such models are particularly important for minimizing service fragmentation and optimizing functional recovery. Effective system models recognize that individuals with depressive disability may require varying levels of care over time. By aligning service intensity with clinical severity and functional impairment, healthcare systems can allocate resources efficiently while maintaining person-centered care. Two widely adopted approaches in this context are the integration of primary, secondary, and tertiary care services and the use of collaborative and stepped-care frameworks.

4.3.1 Primary–Secondary–Tertiary Care Integration

Integration across primary, secondary, and tertiary levels of care detailed in Table 4.2 forms the backbone of effective depressive disability management. Primary care serves as the first point of contact, facilitating early detection, basic treatment, and

Table 4.2 Comparison of primary–secondary–tertiary integration, collaborative care, and stepped-care models

Model type	Level of care	Key features	Advantages	Limitations
Primary–Secondary–Tertiary Integration	All levels	Seamless referral pathways, shared care plans	Continuity, structured care, reduced duplication	Requires well-developed infrastructure
Collaborative Care Model	Primary & Secondary	Multidisciplinary teams, systematic follow-up, outcome monitoring	Improves adherence and functional recovery	Resource-intensive, requires training
Stepped-Care Model	Community & Specialist	Interventions matched to severity, step up if needed	Efficient resource use, scalable	Requires regular assessment, patient engagement

ongoing monitoring of depressive symptoms and functional status (Funk et al., 2008). The integration of mental health services into primary care improves accessibility, reduces stigma, and enables timely intervention, particularly for individuals with mild to moderate depressive disability. Secondary care provides specialized mental health services, including psychiatric evaluation, psychotherapy, and medication management for individuals with more complex or persistent conditions. Tertiary care settings, such as specialized hospitals and rehabilitation centers, address severe, treatment-resistant depression and significant disability requiring intensive multidisciplinary interventions (Institute of Medicine, 2006; Saraceno et al., 2007). Seamless referral pathways and bidirectional communication among these levels are essential to ensure continuity of care and prevent unnecessary delays or duplication of services.

4.3.2 Collaborative Care and Stepped-Care Models

Collaborative care models emphasize shared responsibility among healthcare providers, with clearly defined roles for primary care physicians, mental health specialists, and care managers. These models typically incorporate systematic case identification, evidence-based treatment, regular outcome monitoring, and specialist supervision, all of which contribute to improved clinical and functional outcomes in depressive disorders (Archer et al., 2012; Gilbody et al., 2006; Unützer et al., 2002). Collaborative care has been shown to enhance treatment adherence, patient satisfaction, and long-term disability reduction. Stepped-care models complement collaborative approaches by matching the intensity of intervention to the individual's level of need. Patients begin with low-intensity, cost-effective interventions and progress to more specialized or intensive treatments if required (Bodenheimer et al., 2002; Bower & Gilbody, 2005; Organisation for Economic Co-operation & Development, 2015). This approach supports efficient resource use while ensuring that individuals with greater disability receive appropriate care. Together, collaborative and stepped-care models represent adaptable, evidence-based strategies for integrating services and improving disability outcomes in depressive disorders across diverse healthcare settings.

4.4 Role of Mental Health Professionals in Integrated Systems

Mental health professionals play a central role in the effective functioning of integrated disability care systems for depressive disorders. Table 4.3 incorporates their coordinated involvement ensures that clinical management is aligned with rehabilitation goals, social support needs, and long-term recovery outcomes. In integrated systems, professionals work collaboratively across disciplines and care settings, moving beyond isolated practice toward shared responsibility for patient-centered outcomes (Drake et al., 2004; Rapp & Goscha, 2011; Slade, 2009). Depressive disability often requires sustained engagement with multiple service providers due to its recurrent nature and interaction with psychosocial stressors. Integrated systems enable mental health professionals to contribute their specialized expertise while participating in collective care planning, monitoring, and evaluation. This collaborative approach enhances continuity of care and reduces the risk of fragmented or inconsistent interventions (Alegría et al., 2011; Institute of Medicine, 2006; Kazdin & Blase, 2011; Wagner et al., 2001).

Table 4.3 Roles and responsibilities of psychiatrists, psychologists, social workers, and rehabilitation specialists

Professional	Responsibilities/ Functions	Contribution to Integrated Care	Example Intervention/ Service
Psychiatrist	Diagnosis, medication management, crisis intervention	Clinical leadership, treatment planning	Pharmacotherapy, case review
Psychologist	Psychotherapy, behavioral interventions, assessment	Symptom management, cognitive rehabilitation	CBT, counseling
Social Worker	Psychosocial support, community liaison, resource facilitation	Enhances adherence, addresses social determinants	Home visits, benefit counseling
Rehabilitation Specialist	Functional assessment, vocational training, occupational therapy	Promotes recovery and social participation	Life-skills training, workplace adaptation

4.4.1 *Psychiatrists, Psychologists, Social Workers, and Rehabilitation Specialists*

Psychiatrists are primarily responsible for diagnostic assessment, pharmacological management, and the treatment of complex or severe depressive disorders. Within integrated systems, psychiatrists also provide clinical leadership, specialist consultation, and supervision to other members of the care team, particularly in collaborative and stepped-care models (Burns & Santos, 1995; Unützer et al., 2002). Their role extends to monitoring treatment response and adjusting interventions in line with functional recovery goals. Psychologists contribute evidence-based psychotherapeutic interventions, such as cognitive-behavioral and interpersonal therapies, and play a key role in assessing cognitive and emotional functioning. Social workers address the social determinants of depressive disability, including family dynamics, employment, housing, and access to welfare benefits. Rehabilitation specialists focus on restoring functional capacity through psychosocial rehabilitation, vocational training, and skills development aimed at improving independence and social participation (Becker & Drake, 2003; Bond et al., 2001).

4.4.2 *Interprofessional Communication and Shared Decision-Making*

Effective interprofessional communication is a defining feature of integrated disability care systems. Regular case discussions, shared documentation, and coordinated treatment planning facilitate information exchange among professionals and ensure consistency in care delivery (Buntin et al., 2011; Drake et al., 2004). Such communication reduces duplication of services and enhances the efficiency and quality of care. Shared decision-making further strengthens integrated systems by actively involving individuals with depressive disorders in their care plans. Mental health professionals collaborate with patients and their families to set realistic goals, select appropriate interventions, and monitor progress, thereby promoting autonomy and engagement in the recovery process (Slade, 2009; Slade et al., 2014). Through interprofessional collaboration and shared decision-making, integrated systems support holistic, recovery-oriented care that addresses both the clinical and functional dimensions of depressive disability.

4.5 Community-Based and Institutional Support Systems

Community-based and institutional support systems play a crucial role in the effective integration of disability care for individuals with depressive disorders. These systems extend care beyond clinical settings, addressing psychosocial, functional, and socioeconomic dimensions of disability that are often inadequately managed within hospital-based mental health services. Table 4.4 explains community-oriented approaches emphasize accessibility, continuity of care, and social inclusion, thereby reducing long-term disability and relapse rates associated with depression (Bond et al., 2001; Harvey et al., 2017; World Health Organization, 2010). Community mental health services form the foundation of integrated disability care by delivering early intervention, follow-up care, psychosocial rehabilitation, and family support within local settings. These services facilitate task-sharing models involving community health workers, peer support groups, and primary healthcare providers, which are particularly effective in resource-limited settings. By decentralizing care delivery, community-based systems help overcome barriers related to stigma, geographical inaccessibility, and treatment discontinuity (Druss & Walker, 2011; Jenkins et al., 2011; Vest & Gamm, 2010).

Institutional support systems, including rehabilitation centers, day-care facilities, and supported employment programs, complement community services by addressing functional impairments and promoting reintegration into society. Such institutions focus on vocational training, life-skills development, and cognitive rehabilitation, enabling individuals with depressive disabilities to regain autonomy and social participation. The collaboration between healthcare institutions and social welfare agencies is essential to ensure comprehensive disability support and long-term recovery (Institute of Medicine, 2006; Patel et al., 2015; Thornicroft et al., 2011). Non-governmental organizations (NGOs) and civil society groups play a significant intermediary role in bridging gaps between healthcare systems and community needs. NGOs often provide advocacy, awareness programs, psychosocial counseling, and community outreach, particularly for marginalized populations. Their involvement enhances service coverage and contributes to the sustainability of integrated

Table 4.4 Types of community and institutional support, their services, target population, and outcomes

Type of Support	Key Functions/ Services Provided	Target Population/ Reach	Outcomes/ Evidence
Community Mental Health Services	Early intervention, follow-up care, psychosocial rehabilitation	Patients in local communities	Improved adherence, reduced relapse
NGOs	Advocacy, counseling, outreach, awareness programs	Marginalized or underserved populations	Enhanced service coverage, reduced stigma
Rehabilitation Centers	Vocational training, cognitive and life-skills rehabilitation	Individuals with functional impairments	Increased independence and social participation
Social Welfare Agencies	Financial support, benefits, housing assistance	Patients with socioeconomic constraints	Improved quality of life and access to services

care models by fostering community ownership and policy engagement (Archer et al., 2012; Freeman et al., 2015; Thornicroft et al., 2016). Overall, the integration of community-based and institutional support systems within a coordinated framework is essential for addressing the multidimensional nature of depressive disability. Strengthening linkages among healthcare providers, rehabilitation services, social welfare systems, and community organizations can significantly improve functional outcomes and quality of life for individuals living with depressive disorders.

4.6 Digital and Technological Support in System Integration

The integration of digital technologies into mental health services is increasingly recognized as a pivotal strategy for improving care coordination, accessibility, and continuity for individuals living with depressive disorders (Wykes et al., 2015). Table 4.5 incorporated these innovations not only address geographical and resource limitations but also enhance engagement, monitoring, and treatment personalization, forming a cornerstone of systems-based integrated care (Hilty et al., 2013; Topol, 2019; Torous et al., 2020; Yellowlees et al., 2018).

4.6.1 Telepsychiatry and Remote Mental Health Services

Telepsychiatry has emerged as a critical modality for delivering psychiatric consultations, therapy sessions, and follow-up care remotely, particularly benefiting patients

Table 4.5 Telepsychiatry, mobile health, EHR, AI and their functions, benefits, and challenges

Technology Type	Purpose/Function	Benefits	Challenges/ Limitations
Telepsychiatry	Remote psychiatric consultations and therapy	Accessibility for rural/ underserved areas, continuity of care	Requires stable internet, digital literacy
Mobile Health Applications (mHealth)	Symptom tracking, psychoeducation, CBT modules	Patient empowerment, early detection of relapse	Engagement variability, privacy concerns
Electronic Health Records (EHRs)	Centralized patient data for multidisciplinary teams	Real-time communication, coordinated care	Interoperability issues, data security
AI & Digital Analytics	Predict relapse, monitor outcomes, guide interventions	Personalized care, decision support, improved functional tracking	High technical expertise, ethical considerations

in rural or underserved areas (Hilty et al., 2013; Yellowlees et al., 2018). Studies indicate that telepsychiatry can achieve clinical outcomes comparable to face-to-face interventions, while also improving appointment adherence, reducing travel burden, and increasing patient satisfaction (Hilty et al., 2013; Yellowlees et al., 2018). In addition, telepsychiatry facilitates interprofessional collaboration by enabling virtual case conferences and joint care planning among psychiatrists, psychologists, social workers, and primary care providers (Knaak et al., 2014). This ensures continuity of care for patients transitioning between community, primary, and specialized mental health services (Buntin et al., 2011; Vest & Gamm, 2010).

4.6.2 Mobile Health Applications and Digital Self-Management

Mobile health (mHealth) platforms provide self-management tools, psychoeducation, symptom tracking, and digital cognitive-behavioral therapy (CBT) modules, empowering patients to actively participate in their recovery process (Topol, 2019; Torous et al., 2020). These applications can monitor mood, sleep, and activity patterns, alerting care teams to early signs of relapse. Integration of mHealth data with electronic health records (EHRs) allows clinicians to make data-driven decisions, adjust treatment plans, and provide timely interventions (Buntin et al., 2011; Topol, 2019). Moreover, mobile platforms can support peer networks and virtual support groups, addressing social isolation a key contributor to depressive disability while fostering patient empowerment and engagement in their care trajectory (Hilty et al., 2013; Torous et al., 2020).

4.6.3 Electronic Health Records and Data Integration

Electronic health records (EHRs) are central to enabling coordination across multidisciplinary and intersectoral teams in integrated disability care systems (Topol, 2019; Vest & Gamm, 2010). EHRs allow seamless documentation of clinical assessments, treatment plans, psychosocial evaluations, and functional outcomes, facilitating real-time communication among psychiatrists, psychologists, social workers, and rehabilitation specialists (Buntin et al., 2011; Vest & Gamm, 2010). By integrating patient data across health, social, and rehabilitation services, EHRs reduce duplication of assessments, prevent discontinuities in care, and enhance the overall efficiency of service delivery. In addition, aggregated data can be utilized for quality improvement initiatives, population-level health monitoring, and policy development, strengthening system-level responsiveness to depressive disability (Topol, 2019; Vest & Gamm, 2010).

4.6.4 Benefits of Digital Integration in Disability Care

Digital technologies support several core objectives of integrated disability care systems:

1. **Enhanced accessibility**—Bridging gaps in rural, remote, or underserved regions.
2. **Continuity of care**—Maintaining longitudinal tracking of clinical and functional outcomes.
3. **Patient empowerment**—Promoting self-management, engagement, and adherence.
4. **Interprofessional collaboration**—Facilitating shared decision-making and coordinated care planning.
5. **Data-driven policy and research**—Enabling population health monitoring and evidence-based interventions.

4.6.5 Challenges and Implementation Considerations

Despite their benefits, the adoption of digital technologies in integrated mental health systems faces several challenges. Technical barriers, including limited internet connectivity, lack of standardized software platforms, and insufficient IT infrastructure, can restrict implementation, especially in low- and middle-income countries (Torous et al., 2020; Vest & Gamm, 2010). Patient-related factors, such as digital literacy, privacy concerns, and resistance to technology-based care, also affect utilization and engagement (Hilty et al., 2013; Torous et al., 2020). Moreover, ethical and legal considerations, including data security, confidentiality, and informed consent, require careful attention when integrating digital tools into clinical workflows (Topol, 2019; Vest & Gamm, 2010). To overcome these challenges, successful implementation of digital and technological interventions requires:

- Policy-level support and funding for infrastructure development.
- Training programs for healthcare providers on digital tools and telehealth delivery.
- Standardized protocols for data management, interoperability, and quality assurance.
- Patient education and support to improve digital engagement.

4.6.6 Implications for Integrated Care Systems

Digital and technological interventions hold the potential to transform integrated disability care by:

- Enabling flexible, scalable, and patient-centered care.
- Supporting continuity of services across primary, secondary, and tertiary care levels.

- Facilitating longitudinal monitoring and early intervention for relapse prevention.
- Bridging gaps between clinical, rehabilitative, and community-based services.

As mental health systems evolve, the integration of telepsychiatry, mHealth, and EHRs represents a sustainable and evidence-based strategy to strengthen functional outcomes, improve patient engagement, and enhance the overall quality of disability care in depressive disorders.

4.7 Barriers to Effective Systems Integration

Despite the proven benefits of systems-based integration in depressive disability care, several barriers impede its effective implementation. These challenges span structural, economic, workforce, social, and policy domains and often result in fragmented services, reduced access to care, and suboptimal patient outcomes (Henderson et al., 2013; Saraceno et al., 2007; Saxena et al., 2007). Understanding and addressing these barriers is critical for the successful functioning of integrated care systems (Fig. 4.2).

4.7.1 *Structural Barriers*

Structural barriers refer to limitations inherent in the organization and delivery of health services. Fragmented service delivery, lack of standardized protocols, and poor coordination between primary, secondary, and tertiary care levels disrupt continuity

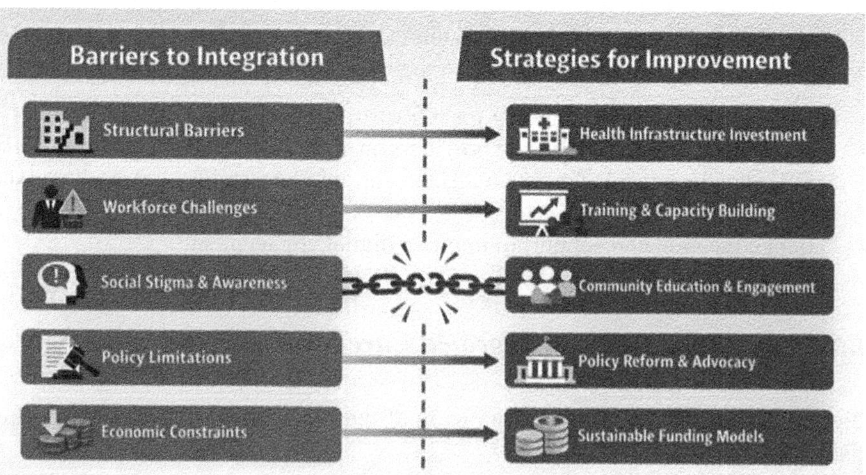

Fig. 4.2 Barriers and strategies in integrated disability care for depressive disorders

of care for individuals with depressive disorders (Lund et al., 2010; Saraceno et al., 2007). Inadequate infrastructure, limited availability of specialized rehabilitation centers, and insufficient integration of community-based services further exacerbate service gaps (Patel et al., 2015; Saxena et al., 2007). These structural deficiencies are particularly pronounced in low-resource settings, where health systems are often overstretched and underdeveloped (Harvey et al., 2017; Vest & Gamm, 2010).

4.7.2 Economic and Resource Constraints

Limited financial resources pose a significant barrier to integrated disability care. Funding constraints restrict the availability of specialized personnel, digital health platforms, and rehabilitation programs (Minas et al., 2012; Saraceno et al., 2007). High out-of-pocket costs for patients can also hinder access to essential services, disproportionately affecting low-income and marginalized populations (Kleinman, 1988; Lund et al., 2010). Additionally, resource limitations impede the implementation of evidence-based interventions, resulting in reduced quality of care and poorer functional outcomes (Henderson et al., 2013; Saraceno et al., 2007).

4.7.3 Workforce Challenges

Integrated care systems rely on a multidisciplinary workforce comprising psychiatrists, psychologists, social workers, rehabilitation specialists, and primary care providers. Shortages of trained mental health professionals, uneven distribution of staff, and high turnover rates present critical barriers to effective collaboration and sustained service delivery (Lund et al., 2010; Patel et al., 2015; Saxena et al., 2007). Moreover, insufficient training in interprofessional communication, care coordination, and digital health utilization limits the capacity of teams to operate cohesively (Buntin et al., 2011; Slade et al., 2014; Vest & Gamm, 2010).

4.7.4 Social and Cultural Barriers

Stigma surrounding mental illness remains a pervasive barrier to system integration. Individuals with depressive disorders may avoid seeking care due to fear of social discrimination, internalized stigma, or lack of family support (Harvey et al., 2017; Lund et al., 2010; World Health Organization, 2010). Cultural beliefs and misconceptions about mental health further reduce engagement with formal healthcare services, limiting the effectiveness of integrated models (Dixon, 2000; Kleinman, 1988). Community-based interventions must therefore address stigma through education, awareness campaigns, and culturally sensitive approaches (Bond et al., 2001; Harvey et al., 2017).

4.7.5 *Policy-Level Limitations*

Policy and regulatory frameworks significantly influence the implementation of integrated care. Inadequate mental health policies, absence of standardized care guidelines, and poor coordination between health and social welfare sectors restrict the ability of systems to deliver comprehensive disability care (Minas et al., 2012; Saraceno et al., 2007). Fragmented funding streams and lack of monitoring mechanisms further hinder accountability and quality assurance (Patel et al., 2015; Saxena et al., 2007). Strengthening policy frameworks, establishing clear operational protocols, and incentivizing intersectoral collaboration are essential to overcome these barriers (Patel et al., 2018; World Health Organization, 2022).

4.7.6 *Implications for Practice*

Addressing these barriers requires a multifaceted strategy:

- Investment in infrastructure and digital technologies to enhance service delivery.
- Capacity building and workforce training in interdisciplinary collaboration and integrated care principles.
- Policy reforms to ensure coordination across healthcare, rehabilitation, and social welfare sectors.
- Community engagement initiatives to reduce stigma and improve access to care.

By systematically targeting structural, economic, workforce, social, and policy barriers, integrated care systems can achieve continuity, accessibility, and quality in disability care for depressive disorders, ultimately improving functional outcomes and quality of life.

4.8 Strategies to Strengthen Integrated Disability Care

To build up the integrated disability care approach for depressed patients, not just one method but a whole assortment of methods shall be needed which consequently eliminate the hurdles that have been pointed out in the preceding sections. First and foremost, the policy transformation measures will always be the footing, legislation transparency, uniform treatment methods, and funding entities that work together guarantee service provision at all levels of care (Saraceno et al., 2007; Saxena et al., 2007; World Health Organization, 2022). The creation of services that demand mutual care and gradual treatment models means that the health systems could give the proper level of treatments to the patients, while at the same time, there would be communication among the different medical specialists and also the support services from the community and social areas would be in sync (Archer et al., 2012; Bodenheimer et al., 2002; Bower & Gilbody, 2005; Gilbody et al., 2006; Unützer et al., 2002). The supply of workforce is utmost factor as well in the making of integrated care successful. Such professional fields like psychiatry, psychology, social work, rehabilitation, and primary care all need to get through training and lifelong continuing education to develop skills in interprofessional collaboration, care coordination, and managing depressive disability through evidence-based practices (Buntin et al., 2011; Slade et al., 2014). With the new technologies such as telepsychiatry and mobile health applications, the providers are able to perform their tasks more effectively and even support patients from far away or less accessible areas, thus, the integrated systems are not limited by their scopes anymore (Buntin et al., 2011; Hilty et al., 2013; Topol, 2019; Torous et al., 2020; Vest & Gamm, 2010; Yellowlees et al., 2018). Mentorship programs, supervision frameworks, and incentives also help to retain staff, maintain team cohesion, and reduce disruptions in care continuity (Lund et al., 2010; Patel et al., 2015; Saxena et al., 2007). Community engagement and awareness initiatives work alongside clinical interventions in different ways; they reduce stigma, make it easier for people to seek help and support, and also aid in the social reintegration of such people. Public education campaigns, peer-support networks, and collaborations with NGOs and social welfare agencies increase the reach and sustainability of community-based rehabilitation programs, particularly in low-resource settings (Archer et al., 2012; Bond et al., 2001; Dixon, 2000; Harvey et al., 2017; Thornicroft et al., 2016; World Health Organization, 2010). By fostering participation, ownership, and culturally sensitive approaches, community involvement strengthens adherence to care plans and ensures that the broader social determinants of depressive disability are addressed (Bond et al., 2001; Harvey et al., 2017; World Health Organization, 2010). Digital technologies are becoming more crucial in the integration of systems. Telepsychiatry, e-health apps, and e-records are some ways to support ongoing monitoring, patient involvement, and communication among doctors, which eliminates patient splitting and enhances the service delivery process (Buntin et al., 2011; Hilty et al., 2013; Topol, 2019; Torous et al., 2020; Vest & Gamm, 2010; Yellowlees et al., 2018). The integration of digital tools allows for real-time tracking of clinical and functional outcomes, facilitates data-driven decision-making, and provides valuable insights for quality improvement and policy development (Minas et al., 2012; Topol, 2019; Vest & Gamm, 2010). Finally, systematic monitoring, evaluation, and research are critical to sustaining and refining integrated care systems. The processes of measuring treatment adherence, functional outcomes, patient satisfaction, and cost-effectiveness have been confirmed as a basis for continual improvement and also as a source for the successful transfer of models to other places (Minas et al., 2012; Topol, 2019; Vest & Gamm, 2010). The study of culturally adapted methods and implementation strategies is most applicable to low- and middle-income countries, where resource limitations exist and the service delivery challenges are most serious (Bodenheimer et al., 2002; Patel et al., 2018; World Health Organization, 2022).

Through policy reform, workforce development, community engagement, digital integration, and evidence-based evaluation, health systems can develop comprehensive, patient-centered, and sustainable integrated disability care programs. Not only do such methods result in superior clinical outcomes but they also enable functional rehabilitation, people's social involvement, and better life quality for those suffering from depression (Bond et al., 2001; Harvey et al., 2017; Saraceno et al., 2007; World Health Organization, 2010, 2022).

4.9 Case Illustrations of Integrated Care Models

Integrated care models for depressive disability have been successfully implemented in several national and international contexts, providing valuable lessons for both high-resource and low-resource settings (Lloyd-Evans et al., 2014). In high-income countries, the Collaborative Care Model (CCM) has been developed in primary care settings at a considerable scale to improve the outcomes for patients suffering from depression and having functional impairment. It puts the spotlight on the coordinated management of primary care doctors, psychiatrists, psychologists, and care managers with systematic follow-up and monitoring for outcomes. The evidence shows that the CCM lowers symptoms of depression, increases patients' adherence to treatment, and supports their recovery, to mention just a few of its benefits, and all this while being cost-effective and powered through even the most basic health infrastructure (Archer et al., 2012; Bodenheimer et al., 2002; Bower & Gilbody, 2005; Gilbody et al., 2006; Unützer et al., 2002). On the other hand, countries like the UK, Netherlands, Australia, and others have taken another route focusing on what they have called the Stepped-Care Model. This is where the weakest patients get low-intensity, evidence-based interventions at the onset, marking the escalation to treatments with higher intensity along a clinically based triage dictated by severity and impairment. Stepped care has proved through practice to be effective by optimizing resource allocation and at the same time making it possible for the specialists to provide care to the most disabled patients. Notably, the model envisions the integration of psychosocial rehabilitation and community support as central in the recovery process, and it makes that very clear through the term "recovery-oriented care" (Bodenheimer et al., 2002; Bower & Gilbody, 2005; Yellowlees et al., 2018). Low- and middle-income countries have come up with innovative ideas that have smoothly integrated the principles of these integrated care concepts even in harsh resource settings. One noteworthy example is India, where the District Mental Health Programme (DMHP) takes full advantage of the primary healthcare infrastructure and provides community-based mental health services, including rehabilitation, social support, and psycho-education under the supervision of district mental health teams. Involving trained non-specialist health workers and community volunteers in the care process has maximized the coverage and maintained the continuity of mental healthcare, while support from NGOs has helped in building strong social support networks. DMHP has been proving through evaluations that it has succeeded in opening-up access to treatment, alleviating depressive symptoms, and producing functional outcomes as good as in the case of disability care integration even in low-resource areas (Bond et al., 2001; Harvey et al., 2017; Vest & Gamm, 2010; World Health Organization, 2010). On the same note, in Brazil, the Family Health Strategy (FHS) takes mental healthcare onboard and embeds it within the purview of primary care teams, thus creating a bridge between clinical services and community and social programs. Patients are the lucky ones as they get seamless care that encompasses all aspects of their illness: pharmacological management, psychotherapy, vocational support, and community rehabilitation. The use of digital health tools such as teleconsultation and mobile monitoring is one of the strategies employed to ensure continuity in care and track functional outcomes (Buntin et al., 2011; Hilty et al., 2013; Topol, 2019; Torous et al., 2020; Vest & Gamm, 2010; Yellowlees et al., 2018). These examples highlight key lessons for implementing integrated disability care systems. First, multidisciplinary collaboration and intersectoral coordination are critical to ensuring that clinical, functional, and social needs are met. Secondly, scalability implies that local resources should be the basis for interventions, community health workers should be used, and task-sharing strategies should be employed. Communication, monitoring, and patient engagement are integrated by digital technologies, thus improving the efficiency and quality of care (Buntin et al., 2011; Hilty et al., 2013; Lund et al., 2010; Topol, 2019; Torous et al., 2020; Vest & Gamm, 2010; Yellowlees et al., 2018). Next, the process of constant evaluation and research will help to smoothen the main path of care, assist in making the right policy choices, and allow the transfer of successful models from one setting to another (Minas et al., 2012; Topol, 2019; World Health Organization, 2022). Real-life examples from various countries indicate that the application of integrated care models has the potential not only to a great extent to cut down the depressive disability, raise up social participation, and make the quality of life better. These models offer the low- and middle-income countries a guiding principle for creating integrated disability care systems that are sustainable, culturally accepted, and appropriate to the resources available (Bond et al., 2001; Harvey et al., 2017; Saraceno et al., 2007; World Health Organization, 2010, 2022).

4.10 Future Directions in Systems-Based Disability Care

The direction of systems-based disability care for people with depressive disorders in the future is expected to be profoundly influenced by technology, service integration, and policy frameworks that make care more accessible, efficient, and patient-centered. The new tendencies are pointing toward precision mental health, which is using data at the individual level, digital monitoring, and predictive analytics to adapt the treatment to the needs and functional goals of each patient (Topol, 2019; Torous et al., 2020; World Health Organization, 2022; Yellowlees et al., 2018). The care pathways that are personalized and include genetic, psychosocial, and environmental information can boost the effectiveness of the treatment, decrease the side effects, and speed up the recovery of functions. The technologies that are coming up such as AI, machine learning, and digital phenotyping will be the major ones in the whole process of predicting depressive relapses, monitoring symptoms' course, and supporting decision-making in care integration systems (Slade et al., 2014; Topol, 2019). Besides, these tools will allow for quick communication between the different specialists in the care team; they will also help increase the adherence to the interventions based on the good evidence and the establishment of the care models that will respond to the changes in patients' functional status over time. The integration of community-based approaches is also going through the process of change, as there is a growing interest in recovery-oriented and socially inclusive ones. It is expected that the collaboration of healthcare systems, NGOs, social welfare agencies, and local communities will become broader with the focus on the components of mental health that are psychosocial rehabilitation, vocational training, and peer support networks (Harvey et al., 2017; World Health Organization, 2010). These are times in low- and middle-income countries where such strategies are especially important, as the lack of resources obliges to share the tasks, engage the communities, and use care models that are culturally adapted to ensure equitable access and sustainability (Archer et al., 2012; Dixon, 2000; Reilly et al., 2013; Thornicroft et al., 2016). Policy development will remain the main factor that decides the future of disability care integration. Coordination of health, social, and educational services; establishment of uniform care protocols; and cross-sector collaboration incentives are some of the main activities prioritized by national mental health strategies that are gradually giving more attention to the issue (World Health Organization, 2022). To keep and enlarge the successful programs, funding models will have to be developed that allow for the integration of digital tools, workforce training, and community-based interventions. Continuous monitoring, evaluation, and research are also essential for maintaining and enlarging integrated systems. Longitudinal data on functional outcomes, cost-effectiveness, and patient satisfaction will be the way to determine system improvements and can also be used to support evidence-based policy decisions (Minas et al., 2012; Topol, 2019). Moreover, international partnerships and knowledge sharing have the potential to adapt the best practices from the high-resource environments to the local settings, thus making the innovations accessible, culturally suitable, and impactful (Harvey et al., 2017; World Health Organization, 2010). The systems-based disability care‘s outlook is digital technologies, precision mental health, community-based services, and policy-driven frameworks all together. If the health systems were to implement these innovations, they would be able to deliver caring that is not only sustainable but also very inclusive and highly centered around the patient, taking into account the clinical as well as the functional aspects of the depressive disorders thus, boosting the quality of life, social participation, and long-term recovery (Torous et al., 2020; World Health Organization, 2010, 2022; Yellowlees et al., 2018).

4.11 Conclusion

Integrated disability care signifies a ground-breaking method in handling depressive disorders by catering to both clinical symptoms and functional limitations through the synchronized and multidisciplinary approach of intersectoral strategies. A systems-based integration guarantees that there is no separation of care regardless of whether it is primary, secondary, or tertiary; besides, it takes advantage of community support, institutional resources, and digital technologies to improve accessibility, efficiency, and patient engagement in the healthcare process. The presence of mental health professionals in the integrated systems is of prime importance, where psychiatrists, psychologists, social workers, and rehabilitation specialists work together to give patient-centered treatment, facilitate shared decision-making, and apply interventions based on scientific evidence. The community-based and institutional support systems comprising of non-governmental organizations, rehabilitation centers, and social welfare agencies play an important role in the area of social determinants by lessening the stigma, thus facilitating integration and independence for those suffering from a depressive disability. In spite of the progress made in this direction, there are still structural, economic, workforce, social, and policy-related barriers that collectively impede effective integration. The establishment of strategies to strengthen the care system such as, for instance, policy reform, service redesign, workforce development, and technological integration, as well as community engagement, is necessary in the quest to develop sustainable and scalable models of care. Innovations in digital health, telepsychiatry, mobile applications, precision mental health, and AI-driven monitoring are among the methods that will still further enhance the capacity in the future of integrated systems to deliver individualized, evidence-based care. The ongoing scrutiny, assessment, and research are going to indicate the amelioration of these models and give the policymakers and practitioners the necessary information, thus making sure that the systems-based disability care is still time, responsive, inclusive, and effective among the different groups and types of medical care. A systems-based integration is the future of disability care in depressive disorders, and it is the one that delivers a holistic, sustainable, and patient-centered approach that treats both clinical and functional aspects of mental health, thus, through the fining of the recovery roads, allowing the participation in the society and the quality of life of the affected individuals to be improved.

References

Alegría, M., Chatterji, P., Wells, K., Cao, Z., Chen, C. N., Takeuchi, D., et al. (2011). Disparities in mental health and mental health care. *Psychiatric Services, 62*(9), 1024–1032.

Archer, J., Bower, P., Gilbody, S., Lovell, K., Richards, D., Gask, L., et al. (2012). Collaborative care for depression and anxiety problems. *The Cochrane Database of Systematic Reviews*, 10, CD006525.

Becker, D. R., & Drake, R. E. (2003). *A working life for people with severe mental illness*. Oxford University Press.
Bhugra, D., Till, A., & Sartorius, N. (2013). What is mental health? *World Psychiatry, 12*(3), 222–231.
Bodenheimer, T., Wagner, E. H., & Grumbach, K. (2002). Improving primary care for patients with chronic illness. *JAMA, 288*(14), 1775–1779.
Bond, G. R., Drake, R. E., Mueser, K. T., & Becker, D. R. (2001). An update on supported employment for people with severe mental illness. *Psychiatric Services, 52*(3), 313–322.
Bower, P., & Gilbody, S. (2005). Stepped care in psychological therapies: Access, effectiveness and efficiency. *British Journal of Psychiatry, 186*, 11–17.
Buntin, M. B., Burke, M. F., Hoaglin, M. C., & Blumenthal, D. (2011). The benefits of health information technology. *Health Affairs, 30*(3), 464–471.
Burns, T., & Santos, A. B. (1995). Assertive community treatment: An update of randomized trials. *British Journal of Psychiatry, 166*(6), 699–706.
Corrigan, P. W., Morris, S. B., Michaels, P. J., Rafacz, J. D., & Rüsch, N. (2012). Challenging the public stigma of mental illness. *American Psychologist, 67*(8), 635–652.
Dixon, L. (2000). Assertive community treatment: Twenty-five years of gold. *Community Mental Health Journal, 36*(1), 1–10.
Drake, R. E., Goldman, H. H., Leff, H. S., Lehman, A. F., Dixon, L., Mueser, K. T., et al. (2004). Implementing evidence-based practices for people with mental illness. *Psychiatric Services, 55*(11), 1213–1222.
Drake, R. E., & Wallach, M. A. (2000). Dual diagnosis: 15 years of progress. *Psychiatric Services, 51*(5), 649–654.
Druss, B. G., & Walker, E. R. (2011). *Mental disorders and medical comorbidity*. Princeton.
Freeman, M.C., Kolappa, K., Caldas de Almeida, J.M., Kleinman, A., Makhashvili, N., Phakathi, S., et al. (2015). Mental health and human rights. *PLoS Medicine, 12*(5), e1001836.
Funk, M., Saraceno, B., Drew, N., & Faydi, E. (2008). *Integrating mental health into primary healthcare*. World Health Organization.
Gilbody, S., Bower, P., Fletcher, J., Richards, D., & Sutton, A. J. (2006). Collaborative care for depression: A cumulative meta-analysis. *Archives of Internal Medicine, 166*(21), 2314–2321.
Harvey, S. B., Modini, M., Joyce, S., Milligan-Saville, J. S., Tan, L., Mykletun, A., et al. (2017). Can work make you mentally ill? *Lancet Psychiatry, 4*(9), 719–727.
Henderson, C., Evans-Lacko, S., Thornicroft, G. (2013). Mental health stigma and discrimination. *British Journal of Psychiatry,* 202(s55), s55–61.
Hilty, D. M., Ferrer, D. C., Parish, M. B., Johnston, B., Callahan, E. J., & Yellowlees, P. M. (2013). The effectiveness of tele-mental health. *Telemed e-Health, 19*(6), 444–454.
Institute of Medicine. (2006). *Improving the quality of health care for mental and substance-use conditions*. National Academies Press.
Jenkins, R., Baingana, F., Ahmad, R., McDaid, D., & Atun, R. (2011). Mental health and the global agenda. *International Journal of Mental Health Systems, 5*(1), 21.
Kazdin, A. E., & Blase, S. L. (2011). Rebooting psychotherapy research. *Perspectives on Psychological Science, 6*(1), 21–37.
Kessler, R. C., Berglund, P., Demler, O., Jin, R., Koretz, D., Merikangas, K. R., et al. (2003). The epidemiology of major depressive disorder. *JAMA, 289*(23), 3095–3105.
Kleinman, A. (1988). *Rethinking psychiatry: From cultural category to personal experience*. Free Press.
Knaak, S., Modgill, G., & Patten, S. B. (2014). Key ingredients of anti-stigma programs. *Healthcare Management Forum, 27*(2), 52–57.
Lloyd-Evans, B., Johnson, S., Slade, M., Barrett, B., Byford, S., Gilburt, H., et al. (2014). Crisis resolution and home treatment teams. *British Journal of Psychiatry, 205*(6), 433–444.
Lund, C., Breen, A., Flisher, A. J., Kakuma, R., Corrigall, J., Joska, J. A., et al. (2010). Poverty and mental disorders. *Social Science and Medicine, 71*(3), 517–528.

Marmot, M., Friel, S., Bell, R., Houweling, T. A. J., & Taylor, S. (2008). Closing the gap in a generation. *Lancet, 372*(9650), 1661–1669.

Minas, H., Lewis, M., Jairam, R., Kakuma, R., Galapatti, A., Ramesh, M., et al. (2012). Mental health system development in low- and middle-income countries. *International Journal of Mental Health Systems, 6*(1), 9.

Murray, C. J. L., & Lopez, A. D. (1996). *The global burden of disease*. Harvard School of Public Health.

Organisation for Economic Co-operation and Development. (2015). *Mental health and work*. OECD Publishing.

Patel, V., Chisholm, D., Dua, T., Laxminarayan, R., & Medina-Mora, M. E. (2015). *Disease control priorities, third edition: Mental, neurological, and substance use disorders*. World Bank.

Patel, V., Flisher, A. J., Hetrick, S., & McGorry, P. (2011). Mental health of young people. *Lancet, 377*(9760), 225–235.

Patel, V., Saxena, S., Lund, C., Thornicroft, G., Baingana, F., Bolton, P., et al. (2018). The Lancet Commission on global mental health and sustainable development. *Lancet, 392*(10157), 1553–1598.

Patel, V., & Thornicroft, G. (2009). Packages of care for mental disorders in low-income countries. *PLoS Medicine, 6*(10), Article e1000160.

Prince, M., Patel, V., Saxena, S., Maj, M., Maselko, J., Phillips, M. R., et al. (2007). No health without mental health. *Lancet, 370*(9590), 859–877.

Rapp, C. A., & Goscha, R. J. (2011). *The strengths model: A recovery-oriented approach to mental health services*. Oxford University Press.

Reilly, S., Newton, L., Bramley, G., Burns, T., Gray, R., McGilloway, S., et al. (2013). Case management approaches for people with severe mental illness. *The Cochrane Database of Systematic Reviews*, 2, CD000050.

Saraceno, B., van Ommeren, M., Batniji, R., Cohen, A., Gureje, O., Mahoney, J., et al. (2007). Barriers to improvement of mental health services. *Lancet, 370*(9593), 1164–1174.

Saxena, S., Thornicroft, G., Knapp, M., & Whiteford, H. (2007). Resources for mental health. *Lancet, 370*(9590), 878–889.

Slade, M., Amering, M., & Oades, L. (2014). Recovery: An international perspective. *World Psychiatry, 13*(1), 12–20.

Slade, M. (2009). *Personal recovery and mental illness: A guide for mental health professionals*. Cambridge University Press.

Thornicroft, G., Alem, A., Antunes Dos Santos, R., Barley, E., Drake, R. E., Gregorio G, et al. (2011). *Community mental health: Putting policy into practice globally*. Wiley-Blackwell.

Thornicroft, G., Mehta, N., Clement, S., Evans-Lacko, S., Doherty, M., Rose, D., et al. (2016). Evidence for effective interventions to reduce mental-health-related stigma. *Lancet, 387*(10023), 1123–1132.

Thornicroft, G. (2006). *Shunned: Discrimination against people with mental illness*. Oxford University Press.

Topol, E. (2019). *Deep medicine: How artificial intelligence can make healthcare human again*. Basic Books.

Torous, J., Jän Myrick, K., Rauseo-Ricupero, N., & Firth, J. (2020). Digital mental health and COVID-19. *Lancet Psychiatry, 7*(8), 651–652.

United Nations. (2006). *Convention on the rights of persons with disabilities*. United Nations.

Unützer, J., Katon, W., Callahan, C. M., Williams, J. W., Jr., Hunkeler, E., Harpole, L., et al. (2002). Collaborative care management of late-life depression in the primary care setting. *JAMA, 288*(22), 2836–2845.

Vest, J. R., & Gamm, L. D. (2010). Health information exchange: Persistent challenges and new strategies. *Health Affairs, 29*(2), 288–294.

Vos, T., Lim, S. S., Abbafati, C., Abbas, K. M., Abbasi, M., Abbasifard, M., et al. (2020). Global prevalence and burden of depressive disorders in 195 countries and territories. *Lancet Psychiatry, 7*(10), 851–864.

Wagner, E. H., Austin, B. T., Davis, C., Hindmarsh, M., Schaefer, J., & Bonomi, A. (2001). Improving chronic illness care: Translating evidence into action. *Health Affairs, 20*(6), 64–78.

Whiteford, H. A., Degenhardt, L., Rehm, J., Baxter, A. J., Ferrari, A. J., Erskine, H. E., et al. (2013). Global burden of disease attributable to mental and substance use disorders. *Lancet, 382*(9904), 1575–1586.

World Health Organization. (2011). *World Bank*. World Health Organization.

World Health Organization. (2010). *Community-based rehabilitation: CBR guidelines*. World Health Organization.

World Health Organization. (2017). *Depression and other common mental disorders: Global health estimates*. World Health Organization.

World Health Organization. (2021). *Guidelines on mental health at work*. World Health Organization.

World Health Organization. (2013). *Mental health action plan 2013–2020*. World Health Organization.

World Health Organization. (2022). *World mental health report: Transforming mental health for all*. World Health Organization.

Wykes, T., Haro, J. M., Belli, S. R., Obradors-Tarragó, C., Arango, C., Ayuso-Mateos, J. L., et al. (2015). Mental health research priorities. *Lancet Psychiatry, 2*(12), 1036–1042.

Yellowlees, P., Shore, J., & Roberts, L. (2018). Practice guidelines for videoconferencing-based telemental health. *American Journal of Psychiatry, 175*(3), 227–234.

Chapter 5
Person-Centered Care Models for Individuals with Chronic Depressive Disability

Ashu Yadav, Ritik Kumar Vaish, and Fariya Khan

Abstract Depression is one of the leading sources of disabilities in the world, affecting more than 300 million people in general. Along with the rising cases of depression, the increasing instances of anxiety disorders also contribute substantially to the overall impact of mental health issues in the global health burden. The World Health Organization identifies the third most prevalent reason for the overall burden of diseases in the world as people with depressive disorder. The condition has become more dismal ever since the COVID-19 pandemic, as people have largely been encountering isolation, economic depression, loss of loved ones, and more psychosomatic pressures. Therefore, the perfect prognosis about the condition has been given by the World Health Organization, which foresees that depression may become the leading reason for the overall burden of diseases in the world by 2030. A number of treatment strategies have been considered in dealing with the effects of depression. These include pharmacologic therapies, psychotherapy, lifestyle interventions, and community-based interventions. Among the treatment paradigms worth considering is the person-centered care approach. This approach has been recognized for its holistic perspective on care. It is patient-centered and has continued to adapt in such a way as to be more in conformance with the behaviors and values exhibited by the patient due to the possible causative nature presented in the situation involving the patient's depression.

Keywords Person-Centered Care · Depression · Disorder · Chronic · Anxiety

A. Yadav
Axis Institute of Higher Education, Kanpur, India

R. K. Vaish
Computational Chemistry Lab, Manipal University, Jaipur, India

F. Khan (✉)
Department of Biotechnology, Kanpur Institute of Technology, Kanpur, India
e-mail: khanfariya.khan@gmail.com

A. Kumar (ed.), *Integrating Disability Care in Major Depressive Disorder Through Neurological and Mental Health Perspectives for Empowerment*,
SpringerBriefs in Modern Perspectives on Disability Research,
https://doi.org/10.1007/978-981-95-8955-5_5

5.1 Introduction

Patient-centered care planning has the promise of really being able to change the way we could handle mental illnesses like the case of anxiety and depression. Care plans integrate the most effective drug treatments together with beneficial approaches to the healthy management of the mind, in addition to encouraging other people in self-care and recovery. Individual differences in personality determine the susceptibility to disease, the course of diseases, and the disease outcome in chronic diseases (Kobau et al., 2010). Chronic diseases are in need of long-term care, monitoring, or management. Chronic illnesses in more developed countries include cardiovascular disease, chronic obstructive pulmonary disease, diabetes, cancer, asthma, arthritis, obesity, and mental illnesses such as depression (Cloninger & Cloninger, 2011). Major Depressive Disorder (MDD) provides an impressive number of manifestations of both physically and psychologically distressing symptoms like fatigue, weight loss, loss of appetite, disturbances in sleep, loss of motivation, difficulties in thinking, and emotional distress with excessive feelings of guilt. The loss of pleasure in activities or the reduced sensation of pleasure is its most prominent symptom (Disease et al., 2018). Depressive illnesses are the most recurrent type of mental illnesses which, in the last count, affected more than 300 million people in the whole world. Due to their recurrent nature, they are disabling. Thus, they significantly affect the affected person's social functioning (Ferenchick et al., 2019). In the year 2008, the World Health Organization (WHO) described the third main cause of the global disease/health burden as being from major depressive disorder. In 2030, it foretold it to be at the top (Malhi & Mann, 2018). Despite its prevalence and impact on the lives of people, the pathophysiology of the disease has always remained unexplained, thus giving no clear diagnosis of the disease. Also, it has not been possible to develop specific medications (Elwy et al., 2020). Various pathophysiological theories have been put across to specifically identify the cause of the disorder. These include theories such as the Hypothalamo-Pituitary Axis disorder, monoamine theories, inflammation, genetic/epigenetic changes, and structural–functional changes in the brain (Giannelli, 2020; Malhi & Mann, 2018). Major depressive disorder is among the major psychological disorders, with its prevalence estimated at 5% of the entire population of adults.

Conclusively, the prevalence and severity of depression and anxiety disorders have significantly increased in the post-COVID-19 era. Indeed, there is a growing body of evidence that has reported high levels of psychological distress, including heightened stress, anxiety, and symptoms of depression, thereby underlining the profound implications of the global health emergency on mental health (Guedes de Pinho et al., 2021). Research further suggests that health education, early detection, and timely delivery of person-centered interventions can go a long way to alleviate economic and social burdens associated with anxiety disorders. Such findings underscore the effectiveness of care planning approaches such as Pathway to Recovery: Standardized programs incorporating elements of the strengths model case management, which enables the person to address those aspects of life over which he or she may exert some influence and at the same time seek and strengthen his or her personal competencies, interests, and functional strengths (Antai-Otong, 2016). Person-centered care, otherwise referred to as person-directed or person-focused care, is generally accepted as a healthcare model that supports the highest possible degree of patient autonomy, collaboration, and shared decision-making. Indeed, it offers a supportive therapeutic collaboration between a person and healthcare professionals so that care reflects personal values, preferences, and goals (Park et al., 2018). PCC has been related to a number of benefits, including raising engagement, empowerment, and self-advocacy for the patients. For the care providers, it brings out reflective practice, enables professional development, and helps caregivers and recipients establish mutually agreed-on treatment goals (Yun & Choi, 2019). Chronic depressive disability refers to sustained and recurring depressive conditions, including persistent depressive disorder otherwise called dysthymia, recurrent major depressive disorder, and treatment-resistant depression (Schramm et al., 2020). These are often interrelated conditions characterized by cycles of partial recovery and relapse over extended periods. The term "chronic depressive disability" has been increasingly used to emphasize the chronic functional impairments and enduring life consequences of these conditions rather than symptom duration per se (Paykel, 2008).

A comprehensive person-centered approach must also incorporate an individual's social relationships, community networks, and cultural context, as these dimensions are integral to identity formation and lived experience. Recognizing individuals within their broader social and cultural environments is essential, as these factors significantly shape treatment expectations, engagement, and outcomes. This consideration is particularly relevant in increasingly multicultural and globalized societies, where migration and cultural diversity add complexity to personal identity and mental healthcare delivery (Bhugra & Becker, 2005).

5.2 Chronic Depressive Disorders: Clinical and Psychosocial Dimensions

Major depressive disorder can be characterized by its symptoms, including the absence of pleasure in activities that were beforehand enjoyable, feeling guilty all the time, problems with concentration, problems related to sleep patterns, and the presence of suicide attempts/thoughts. Put simply, the mental illness results in mood problems like sadness, feelings of having no purpose, irritability, alongside issues such as the reduced self-feeling and the lack of focus. Physical symptoms include fatigue and low energy levels. According to the Diagnostic and Statistical Manual of Mental Disorders, Fifth Edition (DSM-5), depressive disorders fall into seven categories: major depressive disorder, persistent depressive disorder (dysthymia), premenstrual dysphoric disorder, substance- or medication-induced depressive disorder, disruptive mood dysregulation disorder, other specified depressive disorder, and unspecified depressive disorder (McHugh & Weiss, 2019). Major depressive disorder, dysthymia, and substance-related depressive disorder are the most studied, both in the general population and among people with alcohol use disorder (AUD).

Major depressive disorder is present when there are at least five depressive symptoms for two consecutive weeks, including a depressed mood or the loss of interest or pleasure. Common symptoms may also include feelings of death or suicidal thoughts, changes in appetite or sleep, agitation or feeling slowed down, feeling tired or fatigued, difficulty concentrating or making decisions, feelings of guilt or worthlessness, or excessive self-blame. In comparison, dysthymic disorder, also known as persistent depressive disorder, is less severe but more chronic. It is characterized by at least two years of depressed mood with at least two symptoms that include feelings of hopelessness, changes in appetite or sleep, feeling down, poor self-esteem, difficulty concentrating or making decisions, or feeling indecisive, as outlined by McHugh and Weiss (2019). Alcohol-induced depressive disorder is similar to depression in that it only exists in terms of alcohol consumption or withdrawal, lasting for at least three to four weeks of abstinence (Fig. 5.1).

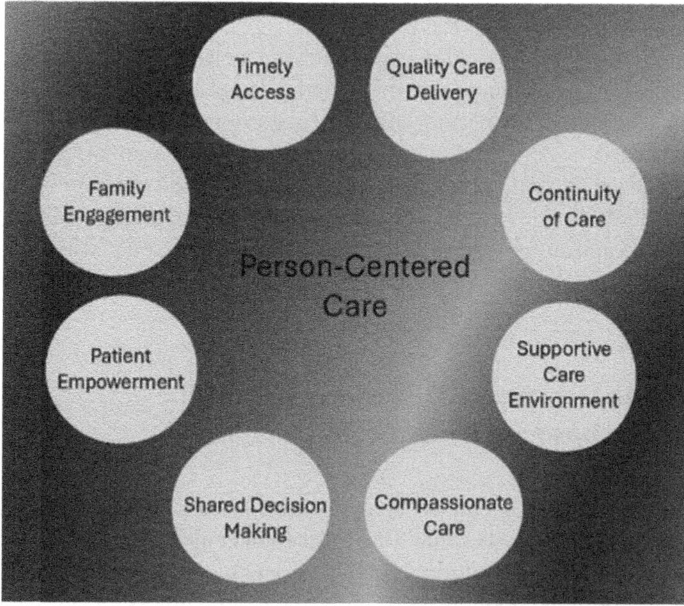

Fig. 5.1 Essential principles informing the person-centered care model

5.3 Burden of Depression and Disability

Depressive illnesses continue to cause significant impairment in individuals, even when treatment is deemed successful and clinical remission has been achieved. In many instances, remission does not imply the full recovery of all depressive symptoms. Residual symptoms remain, often manifesting as cognitive impairments, decreased concentration, problems with memory, or continued social and occupational dysfunction (Baker et al., 2001). These residual symptoms may cause significant distress and impairment in daily functioning, which restricts an individual's full return to premorbid levels of functioning and impinges on the personal, professional, and social spheres of life. Furthermore, the ongoing risk of recurrence and relapse continues to be one of the major challenges in the long-term management of depressive disorders. The persistent uncertainty of symptoms returning can itself adversely impact psychological well-being and quality of life. Longitudinal studies emphasize the chronic course and recurrence of depression. For example, the study by Hardeveld et al. (2009) showed that the recurrence rates of major depressive disorder were substantially higher among individuals treated in specialist mental health settings, with recurrence occurring in 60% of patients within 5 years, 67% within 10 years, and as high as 85% within 15 years. By contrast, recurrence rates in primary care populations were considerably lower, with approximately 35% experiencing recurrence within 15 years. Other evidence suggests that certain factors significantly elevate the risk of recurrence. Previous research has identified the number of earlier episodes of depression and the presence of subclinical residual symptoms as the most powerful predictors of relapse (Doesschate et al., 2010). These findings again stress the need for long-term, person-centered management strategies extending beyond symptom remission to address residual impairments and minimize the risk of future episodes of depression.

5.4 Person-Centered Care Approaches

Initial research into person-centered care has found that there are identifiable, tangible advantages. There is presently a broad literature that identifies practices related to person-centered care with improvement in quality of life, reduction in agitation, improved sleep patterns, and documented elevations in patients' levels of self-esteem. Collectively, these literature sources contend that practices related to person-centered care and its accompanying culture work efficaciously across varied settings of care (Koren, 2010). Perhaps more importantly, it is important for PCC not to be regarded solely, or at least exclusively, as concerning itself with broad strategies toward increasing appropriate care outcomes, care-given communication patterns, improved care staffing ratios, or similar factors that can be provided by varied clinical formats, varied organizational patterns, or assorted interdisciplinary patterns (Table 5.1). In terms of managing chronic depressive disability, practices related to person-centered

Table 5.1 Roles of stakeholders in person-centered care

Stakeholder	Role in PCC	Contribution to outcomes
Individual	Goal setting, self-management	Empowerment
Clinician	Facilitator, guide	Therapeutic alliance
Family/Caregivers	Emotional and practical support	Continuity
Mental health team	Coordinated intervention	Comprehensive care
Community services	Social inclusion	Functional recovery

care strategies are highly salient concerning bridging practical needs between patients needing focused, long-term management for their care issues that encompass not merely temporal or definite episodes oriented or related toward focused, acute mental-psychiatric concerns, but also multidimensional levels for facilitating improved mental, cognitive, functional, or social levels related toward enhancing recovery from depressive disabilities (Sjögren et al., 2013). Person-centered care practices encompass an overarching concern concerning regarding clients' subjective concepts toward their personal ideas, feelings, or concerns regarding personal illnesses or mental disability concerns, or regarding individual patient needs, concerns, or expectations. PCC allows for collaborative decision-making, giving patients an equal role in making decisions about their care and exhorting them to take more control over their lives (Wilberforce et al., 2016). Despite having almost two decades to develop passion for PCC, making PCC a best practice still poses challenges. Such challenges include the amorphous, ambiguous quality of PCC policy documents that can mean different things to different people, to which are added overall system constraints posed by rapidly changing healthcare delivery priorities. Mix this with health care workers who have not fully bought into PCC, along with healthcare administrators who lack preparedness to fully adopt PCC owing to Calls to address other priority healthcare concerns (Lefurgey et al., 2025).

5.5 Person-Centered Mental Healthcare Policies and Guidelines

The movement to embed Person-Centered Care within mental health care is right at the center of current policy and guideline development. The aim couldn't be simpler: to lift the experience of care and the effectiveness of that care for the people using the service. At the most fundamental level, person-centered care involves keeping the care in line with what the users want, what they value, and what they try to achieve, and ensuring that this approach accompanies them through the entire care system, from the hospital, through the community, so that the care is seamless. That which matters most to patients is captured through their personal reports of how they're getting on. However, the value of person-centered care is best illuminated in the context of more difficult or chronic issues, where the establishment of trust and a primary commitment to overall health and well-being is paramount. Harding et al. (2015) capture the central concepts underlying person-centered care for health and care professionals as follows: a set of very practical "points to remember" that include treating the person with kindness, dignity, and respect; recognizing them as individuals, not just as a diagnosis; communicating with them in a way that respects their individual communication choices and cultural differences; respecting their privacy and making the process, not just the goal, paramount; organizing around the needs and goals of the patient, not the organization's internal structure; and integrating the organization's "hospital-centered" approach with "community health" in order to better serve patients' needs through increased coordination and continuity. Effectively, person-centered care enables individualized assistance and treatment that adjusts approaches to a person's history, preferences, assets, and lifestyle proclivities, getting beyond the prescription to work with the individual's creative plan-making. Central to this approach is the enabling and empowerment of individuals, fostering confidence, skills, and autonomy so that they can actively manage their health, pursue valued roles and relationships, and realize their full potential (Pulvirenti et al., 2014).

In all likelihood, the common ground between the principles of person-centered care and those of healthcare policy has significance. We've never wanted to disagree about incorporating what the customer has to say, even if we disagree about how we do the research and incorporate the feedback- and we've changed approaches based on advancements in technology. Technology helps make incorporating what the clients say about their own care and the tools we provide for them even simpler. One of the common threads in several guidelines, such as that developed by the Institute for Patient and Family-Centered Care (Lefurgey et al., 2025), is personalizing the plan of care according to what the client has to say and their own plans.

5.6 Core Components of Person-Centered Care

There are several aspects that are addressed under the umbrella concept of person-centered care, of which five are universally regarded as essential: rights; the relationship between practice and the organization of services; relevance to practitioners, values, or the evidence base; and education. Of these, the rights aspect in mental health practice is particularly essential, since psychiatric practices under consideration can entail the infringement of the individual's liberty in the interests of improved outcomes. While this infringement can be deemed valid from the professional perspective within the context of the individual's ill phase, from the professional's view, the individual needs to be addressed primarily in terms of the concept of liberty alongside fundamental human rights. The balance that needs to be drawn here, therefore, between the mental ill individual's right to live the life he or she wants, values, and pursues, and the imperative of effective treatment, remains particularly essential. On a more debatable shore, this approach to mental health, together with the larger social context within which this practice can exist, can also allow for the critique and assessment of mental health practice, alongside the larger social, economic frameworks within which practice exists. Truly therefore, the person-centered approach to mental health practice, therefore, relies for its success on the individual's awareness of basic human rights and the relevant legislation, the capacity to apply this legislation to differentiate individual freedom only within proportional, ethical boundaries, alongside the professional mindset that prioritizes individual choice, preference, dignity, and rights above the rest, under the leadership of someone like Dinesh Bhugra, who wrote, "A person-centered approach prioritizes the individual's autonomy, dignity, rights, needs, choices, empowerment, individuality, personhood, individuosity, self-determination, self-actualization, self-authorship, self-power" (Bhugra, 2016).

The relationship between practice and service organizational principles forms the central part of realizing success through person-centered care. Personal rehabilitation and principles of a person-centered approach share key values, assumptions, and implications for mental health services. Enhancing a more person-centered approach to mental health services helps to support a recovery-oriented framework to achieve individualized outcomes of recovery (Mezzich et al., 2016). Presently, current practice on diagnosis, therapy, and support becomes increasingly informed by a person's life history, personal strengths, personal goals, social context, activities, personal values, and personal beliefs. This enhances new practice from just addressing symptom relief to questioning how effectively any diagnosis, therapy, or support contributes to a person's personal achievement of his or her personal life goals and living his or her valued lives. Practitioners, including general psychiatrists, are therefore expected to work together on therapy to address patient preferences, despite such preferences not necessarily supporting current levels of existing evidence.

A person-centered approach is important not only from the perspective of patients but also from the perspective of professionals in the sector. Apart from the improvements in the quality of treatment and the satisfaction of patients, the approach can allow improvements in the job satisfaction of professionals and can contribute to the reduction of job-related stress in the sector. Professionals in the healthcare sector feel motivated by the need to help patients strengthen their abilities and resources and participate in meaningful activities of their lives while treating the recipients of the services with dignity and respect in addition to addressing emotional needs and basic needs in the process. Professionals feel that they should practice services in well-integrated environments too. The concept of the person-centered approach considers the fact that professionals are people too, with their own needs and limitations in the process of delivering services in the sector. A professional's well-being is important in the delivery of quality services in the sector because issues of "burnout and depersonalization" can undermine the delivery of caring and competent services in the sector, especially in the current context of recruitment and retention of resources in the sector of psychiatry (Ahmad et al., 2014).

A growing trend in evidence-based medicine is the ease with which professionals adopt guidelines and best practices. This is about navigating through the values which conflict between those who are involved—patients, caregivers, professionals, and organizations. To provide decision-support which is focused on the individual, we need such a collaborative approach which looks at evidence-based medicine as the partner for values-based practice. By basing the essential ethics in clinical practice on the best evidence we have from research, we will have a robust way of practicing. According to the Royal College of Psychiatrists (RCPsych), the eight key values include: Communication, Decency, Empathy, Fairness, Honesty, Humility, Respect, and Trust (George et al., 2015).

A recently done review of training in the curriculum by the Royal College of Psychiatry found that the current curriculum is incomplete in that the skills underlying "person-centered care" are inadequately articulated in the current curriculum. This review recommends that the revised curriculum clearly reflect the strong emphasis on person-centered care articulated in the foreword. Training programs should equip learners with practical skills central to this approach, including cooperative working and collaborative planning of support (Table 5.2). In addition, the curriculum should draw on the broader domain of person-centered care by incorporating content related to ethics, community engagement, social integration, individual rights, and relevant legal frameworks.

Table 5.2 Key dimensions of person-centered approaches in mental health care

Dimension	Core focus	Key elements and implications	References
Rights-Based Perspective	Protection of individual freedom and human rights	Balances the need for effective psychiatric treatment with respect for autonomy, dignity, and legal rights; emphasizes proportional and ethical use of restrictive interventions; enables critique of mental health practices and broader social systems	Bhugra (2016)
Practice and Service Organization Relationship	Integration of person-centered values into service delivery	Aligns personal rehabilitation with recovery-oriented care; shifts focus from symptom control to enabling individuals to pursue meaningful life goals; considers personal history, strengths, values, and social context in care planning	Mezzich et al. (2016)
Relevance for Clinicians	Impact on professional satisfaction and care quality	Enhances clinician motivation, job satisfaction, and well-being; reduces burnout and depersonalization; recognizes clinicians as individuals within supportive service systems; improves recruitment and retention in psychiatry	Ahmad et al. (2014)

(continued)

Table 5.2 (continued)

Dimension	Core focus	Key elements and implications	References
Values and Evidence	Integration of evidence-based and values-based practice	Encourages collaborative clinical decision-making; adapts guidelines to individual circumstances; resolves value conflicts among patients, caregivers, clinicians, and organizations; guided by principles such as empathy, respect, trust, fairness, and honesty	George et al. (2015); RCPsych
Training and Education	Development of person-centered competencies	Calls for explicit inclusion of person-centered skills in psychiatric curricula; emphasizes relational competencies, ethics, human rights, and community engagement; involves patients and caregivers in training; requires formative and summative assessment of person-centered abilities	Royal et al. (2018)

References

Ahmad, N., Ellins, J., Krelle, H., & Lawrie, M. (2014). *Person-centred care: From ideas to action* (pp. 1–100). Health Foundation Google scholar.

Antai-Otong, D. (2016). Caring for the patient with an anxiety disorder. *Nursing Clinics of North America, 51*(2), 173–183. https://doi.org/10.1016/J.CNUR.2016.01.003

Baker, A. (2001). Crossing the quality Chasm: A new health system for the 21st century. *BMJ, 323*, 1192. https://doi.org/10.1136/bmj.323.7322.1192

Bhugra, D. (2016). Bill of rights for persons with mental illness. *International Review of Psychiatry, 28*(4), 335–335. https://doi.org/10.1080/09540261.2016.1210580

Bhugra, D., & Becker, M. A. (2005). Migration, cultural bereavement and cultural identity. *World Psychiatry, 4*(1), 18–24. https://www.ncbi.nlm.nih.gov/pmc/articles/PMC1414713/

Catalyst, N. (2017, January1). *What Is patient-centered care? NEJM Catalyst*. Accessed October 20, 2022. https://doi.org/10.1056/CAT.17.0559

Cloninger, C. R., & Cloninger, K. M. (2011). Development of instruments and evaluative procedures on contributors to illness and health. *The International Journal of Person-Centered Medicine, 1*(3), 456–459. https://doi.org/10.5750/IJPCM.V1I3.99

Constand, M. K., MacDermid, J. C., Dal Bello-Haas, V., & Law, M. (2014). Scoping review of patient-centered care approaches in healthcare. *BMC Health Services Research, 14*, 2–9. https://doi.org/10.1186/1472-6963-14-271

Disease, G., Monasta, L., Ronfani, L., Beghi, E., Giussani, G., Bikbov, B., Perico, N., Bosetti, C., Cortinovis, M., Gallus, S., & Remuzzi, G. (2018). Global, regional, and national incidence, prevalence, and years lived with disability for 354 diseases and injuries for195 countries and territories, 1990–2017: A systematic analysis for the Global Burden of Disease Study 2017. *The Lancet, 392*(10159), 1789–1858. https://doi.org/10.1016/S0140-6736(18)32279-7

Doesschate, M. C., Bockting, C. L. H., Koeter, M. W. J., & Schene, A. H. (2010). Prediction of recurrence in recurrent depression: A 5.5-year prospective study. *The Journal of Clinical Psychiatry, 71*(8), 984–991. https://doi.org/10.4088/JCP.08M04858BLU

Elwy, A. R., Elwy, A. R., Taylor, S. L., Zhao, S., McGowan, M., Plumb, D. N., Westleigh, W., Gaj, L., Yan, G. W., Bokhour, B. G., & Bokhour, B. G. (2020). Participating in Complementary and

Integrative Health Approaches Is Associated With Veterans' Patient-reported Outcomes Over Time. *Medical Care*, 58. https://doi.org/10.1097/MLR.0000000000001357

Ferenchick, E. K., Ramanuj, P., & Pincus, H. A. (2019). Depression in primary care: Part 1-screening and diagnosis. *BMJ (Clinical Research Ed.), 365*, Article l794. https://doi.org/10.1136/bmj.l794

George, R. E., Dogra, N., & Fulford, B. (2015). Values and ethics in mental-health education and training: A different perspective. *The Journal of Mental Health Training, Education and Practice, 10*(3), 189–204. https://doi.org/10.1108/JMHTEP-08-2014-0024

Giannelli, F. R. (2020). Major depressive disorder. *JAAPA: Official Journal of the American Academy of Physician Assistants, 33*(4), 19–20. https://doi.org/10.1097/01.JAA.0000657208.70820.AB

Hardeveld, F., Spijker, J., de Graaf, R., Nolen, W. A., & Beekman, A. T. F. (2009). Prevalence and predictors of recurrence of major depressive disorder in the adult population. *Acta Psychiatrica Scandinavica, 122*(3), 184–191. https://doi.org/10.1111/J.1600-0447.2009.01519.X

Harding, E., Wait, S., Scrutton, J. (2015). *The State of Play in Person-Centred Care: A Pragmatic Review of How Person-Centred Care Is Defined, Applied and Measured, Featuring Selected Key Contributors and Case Studies Across the Field.* The Health Policy Partnership: London, UK. Available online https://www.healthpolicypartnership.com/app/uploads/The-state-of-play-in-person-centred-care.pdf

Kivelitz, L., Schulz, H., Melchior, H., & Watzke, B. (2015). Effectiveness of case management-based aftercare coordination by phone for patients with depressive and anxiety disorders: Study protocol for a randomized controlled trial. *BMC Psychiatry, 15*, 2–7. https://doi.org/10.1186/s12888-015-0469-y

Kobau, R., Sniezek, J. E., Zack, M. M., Lucas, R. E., & Burns, A. (2010). Well-being assessment: An evaluation of well-being scales for public health and population estimates of well-being among US adults. *Applied Psychology: Health and Well-Being, 2*(3), 272–297. https://doi.org/10.1111/J.1758-0854.2010.01035.X

Koren, M. J. (2010). Person-centered care for nursing home residents: The culture-change movement. *Health Affairs, 29*(2), 312–317. https://doi.org/10.1377/HLTHAFF.2009.0966

Lefurgey, S., Detillieux, S., Shaheen, A., Daigle, P., Nolan, D., & Rudnick, A. (2025). Person-centered care: Learning from the evolution of mental health care. *Encyclopedia.* https://doi.org/10.3390/encyclopedia5010029

Malhi, G. S., & Mann, J. J. (2018). Depression. *Lancet (London, England), 392*(10161), 2299–2312. https://doi.org/10.1016/S0140-6736(18)31948-2

McHugh, R., & Weiss, R. (2019). Alcohol use disorder and depressive disorders. *Alcohol Res, 40*(1), arcr.v40.1.01. https://doi.org/10.35946/arcr.v40.1.01

Mead, N., & Bower, P. (2000). Patient-centredness: A conceptual framework and review of the empirical literature. *Social Science and Medicine, 51*, 1087–1110. https://doi.org/10.1016/S0277-9536(00)00098-8

Mezzich, J. E., Botbol, M., Christodoulou, G. N., Cloninger, C. R., & Salloum, I. M. (2016). Introduction to person-centered. *Psychiatry*. https://doi.org/10.1007/978-3-319-39724-5_1

Park, M., Giap, T. T. T., Lee, M., Jeong, H. Y., Jeong, M., & Go, Y. (2018). Patient- and family-centered care interventions for improving the quality of health care: A review of systematic reviews. *International Journal of Nursing Studies, 87*, 69–83. https://doi.org/10.1016/J.IJNURSTU.2018.07.006

Paykel, E. (2008). Partial remission, residual symptoms, and relapse in depression. *Dialogues ClinNeurosci, 10*(4), 431–437. https://doi.org/10.31887/DCNS.2008.10.4/espaykel

Person-Centred Training and Curriculum Scoping Group. (2018). *Person-Centred Care: Implications for Training in Psychiatry (College Report CR215).* Royal College of Psychiatrists, Google scholar.

Pinho, L., Correia, T., Sampaio, F., Sequeira, C., Teixeira, L., Lopes, M., & Fonseca, C. (2021). The use of mental health promotion strategies by nurses to reduce anxiety, stress, and depression during the COVID-19 outbreak: A prospective cohort study. *Environmental Research, 195*, Article 110828. https://doi.org/10.1016/j.envres.2021.110828

Pulvirenti, M., McMillan, J., & Lawn, S. (2014). Empowerment, patient centred care and self-management. *Health Expectations, 17*(3), 303–310. https://doi.org/10.1111/J.1369-7625.2011.00757.X

Schramm, E., Klein, D. N., Elsaesser, M., Furukawa, T. A., & Domschke, K. (2020). Review of dysthymia and persistent depressive disorder: History, correlates, and clinical implications. *The Lancet Psychiatry, 7*(9), 801–812. https://doi.org/10.1016/S2215-0366(20)30099-7

Sjögren, K., Lindkvist, M., Sandman, P.-O., Zingmark, K., Edvardsson, D., & Edvardsson, D. (2013). Person-centredness and its association with resident well-being in dementia care units. *Journal of Advanced Nursing, 69*(10), 2196–2206. https://doi.org/10.1111/JAN.12085

Sullivan-Taylor, P., Suter, E., Laxton, S., Oelke, N. D., & Park, E. (2022). Integrated People-Centred Care in Canada—Policies, Standards, and Implementation Tools to Improve Outcomes. *International Journal of Integrated Care, 22*(1). https://doi.org/10.5334/ijic.5943

Švab, I., & Cerovečki, V. (2024). Person-centred care, a core concept of family medicine. *European Journal of General Practice, 30*(1). https://doi.org/10.1080/13814788.2024.2393860

Wilberforce, M., Challis, D., Davies, L., Kelly, M., Roberts, C., & Loynes, N. (2016). Person-centredness in the care of older adults: A systematic review of questionnaire-based scales and their measurement properties. *BMC Geriatrics, 16*(1), 63. https://doi.org/10.1186/S12877-016-0229-Y

Yun, D., & Choi, J. (2019). Person-centered rehabilitation care and outcomes: A systematic literature review. *International Journal of Nursing Studies, 93*, 74–83. https://doi.org/10.1016/J.IJNURSTU.2019.02.012

Chapter 6
Empowering Individuals Living with Severe Depressive Disorder

Prakash Kumar and Ajay Kumar

Abstract Severe depressive disorder is much larger than just having feelings of sadness at a moment in time; it impacts a person's entire life for the long term by creating changes in how a person's brain functions and processes emotions and logic and creates an inability for the individual to experience happiness. This creates a difficult daily life due to biological changes that are made at an early age; doing basic tasks that are usually completed daily, such as going out in public, participating in group activities with other people, and communicating and meeting other people, becomes practically impossible. In addition to the biological effects of the illness itself, a person living with this illness experiences many very real obstacles, such as poverty, social exclusion, and stigma, that can push them deeper into a cycle of illness by creating additional barriers, making it even harder for them to obtain treatment or get help. Care needs to go beyond treating the physical symptoms associated with depression so that a person can break the cycle of illness. The best way to break this cycle is through early intervention, active treatment by physicians, families, and community support systems, to help empower a person through recognizing what a person is able to do. Addressing all of person's needs quickly and eliminating any social barriers that may hinder a person's independence would allow that person to continue being independent and live the most fulfilling life possible. Today's new approach to treating depression is directed towards creating an environment that is a place where people who are experiencing depression are treated with dignity and feel safe, creating a climate for growth and achieving a sense of empowerment in sustaining hope for a continuous recovery and resiliency.

Keywords Depressive disorder · Empowerment · Neurobiological · Psychosocial · Rehabilitation · Stigma

---

P. Kumar
Department of Biotechnology, ASET, Sharda University, Agra 282007, India

A. Kumar (✉)
Department of Biotechnology, Rama University, Kanpur 209217, India
e-mail: ajaymtech@gmail.com

A. Kumar (ed.), *Integrating Disability Care in Major Depressive Disorder Through Neurological and Mental Health Perspectives for Empowerment*,
SpringerBriefs in Modern Perspectives on Disability Research,
https://doi.org/10.1007/978-981-95-8955-5_6

6.1 Introduction and Conceptual Framework

Empowerment in severe major depressive disorder (MDD) reduces symptom severity and enhances clinical participation through self-management techniques, patient-provider shared decision-making, and supportive social networks and communities (Wang et al., 2022; Zoun et al., 2019). Studies from 2025 reveal that improving self-esteem is a major obstacle to overcoming self-stigma (Smith et al., 2025). Recovery models, particularly patient-centred recovery models, are now focusing on independence and peer support, shifting the emphasis from managing a disability to supporting personal agency and building a social network for recovery.

6.1.1 Overview of Severe Depressive Disorder

Severe depression is a deadening presence that converts every day into a battle against despair, therefore causing your mind to be enveloped in cloudiness and making it so difficult to generate even one idea. It removes all joy from everything that, before, gave life purpose and satisfaction (Chen et al., 2025a, 2025b). Physical changes, including persistent inflammation and brain circuit abnormalities, often cause this condition and might make it challenging to treat with traditional approaches. Modern research sees serious depression as a life-course disease needing long-term, multidimensional treatment plans rather than as an episodic one (Malhi & Mann, 2018).

6.1.2 Depression as a Disabling Condition

Depression is among the most pressing health problems globally since it profoundly affects a person's ability to live, interact socially, and carry out daily tasks, thus contributing significantly to the global disability burden (GBD 2019 Mental Disorders Collaborators, 2022). Many individuals continue to have ongoing difficulties with decision-making and motivation even as the most severe symptoms start to reduce (Lam et al., 2014). For individuals already managing other chronic physical diseases, this burden is bigger; therefore, we have to give first consideration to improving a person's general quality of life and long-term recovery instead of only treating symptoms (Voinov et al., 2023).

6.1.3 *Importance of Empowerment and Integrated Care*

Giving people the capacity to help their own therapy and decisions is a major element in conquering major depression (Hansson & Björkman, 2016). Studies show that this feeling of agency motivates people to follow their course of treatment, hence enhancing their quality of life, and helps reduce the severity of the illness (Tambuyzer et al., 2014). This empowerment is even more effective when mental and physical health services work as one team to provide perfect treatment that leads to markedly better long-term well-being (Tops et al., 2024).

6.2 Neurological and Mental Health Perspectives

Neurological and mental health perspectives emphasize functional brain network disruptions, altered reward processing, executive dysfunction, and impaired hippocampal activity in severe major depressive disorder (MDD), highlighting neurobiology's role in symptomatology and recovery (Tian et al., 2024). Dysregulated neuroplasticity and reduced brain-derived neurotrophic factor (BDNF) are linked to mood disturbances and cognitive deficits (Liu et al., 2025). Integrating neuroscience into mental healthcare supports personalized interventions targeting brain-based mechanisms to empower individuals with severe depression (Liu et al., 2025). Neurological and Mental Health Perspectives are illustrated in Fig. 6.1 for Depressive Disorder.

6.2.1 *Neurological Basis of Severe Depression*

In major depressive disorder, the brain's wiring for pleasure and concentration usually underperforms relative to that of healthy people (Ma et al., 2024). Recent machine learning investigations have shown that depression manifests in unique neural patterns, rather than simply being a result of a chemical imbalance (Chen et al., 2025a, 2025b). MDD essentially reshapes how the brain regulates stress and forms the cellular connections necessary for emotional resilience (Cui et al., 2024).

6.2.2 *Psychological and Emotional Functioning*

Chronic negative thought patterns and damaged emotional control systems linked with major depression lower both emotional and intellectual performance. They usually cause people to feel sadness and intense rumination because their emotional functioning has been damaged by the way they construct obstacles to feeling the

Fig. 6.1 Illustrate neurological and mental health perspectives of depressive disorder

warmth of good experiences by raising the negativity of unpleasant ones (Ahorsu & Tsang, 2017). This change in emotional processing also results in lower emotional resilience and the inability to relate to other people (Ahorsu & Tsang, 2017). The difficulty with memory and focus connected with these cognitive problems immediately causes a chain reaction that negatively impacts every aspect of a person's everyday life and greatly reduces their general quality of living (Villalobos et al., 2021).

6.2.3 Interaction Between Brain, Behaviour, and Disability

Severe depression is defined by a consistent brain-action interplay; alterations in brain networks can cause social withdrawal and compromised decision-making, therefore exacerbating the biological condition (Ma et al., 2024; Cui et al., 2024). This cycle can turn into a loop such that the brain's failure to appropriately process emotions fuels behaviours like overthinking, therefore further cementing the depression. Studies done in 2025 using cutting-edge neural mapping show that how our brains handle ideas about ourselves correlates with the severity of the symptoms we manifest in the actual world (Kommineni et al., 2025). Eventually, this shows that rather than being merely a mental disease or behaviour, depression is a physical and emotional cycle that must be treated concurrently for the brain and the patient's daily life (Cui et al., 2024; Kommineni et al., 2025).

6.3 Functional Impact and Lived Experience

Severe depressive disorder often makes it impossible to work or stay connected, and these difficulties sometimes last even after the primary symptoms pass (Lai et al., 2025; Luo et al., 2024). Those living with it often describe a painful loss of control and a heavy weight of stigma that makes navigating everyday life an exhausting struggle (Lai et al., 2025). The Functional Impact and Lived Experience are summarized in Fig. 6.2 for Severe Depression.

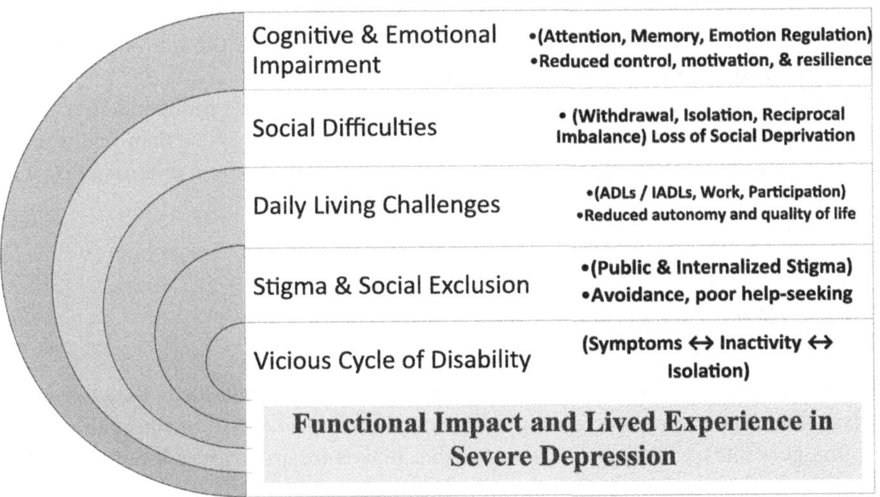

Fig. 6.2 Showing functional impact and lived experience in severe depression

6.3.1 Cognitive, Emotional, and Social Impairments

The chaos that major sadness can cause in the cognitive and interpersonal skills we use daily is only one aspect of a person's mood. People who struggle with their memory, concentration, and capacity for basic decisions may find it difficult to live alone—a burden especially heavy for the elderly (Wang et al., 2025; Zhou et al., 2020). The emotional load of negative ideas and lack of happiness might drive people to withdraw from their loved ones, causing a gloomy cycle of isolation (Zhou et al., 2020). Because of the challenge of maintaining relationships and social status under such circumstances, the impairment brought about by depression grows into a very close, all-consuming struggle (Zhou et al., 2020).

6.3.2 Daily Living and Participation Challenges

Considering the most basic elements of daily life, financial management, housekeeping, and even travel might seem like impassable hurdles for someone with severe depression. According to recent studies from 2024 and 2025, there is a vicious circle: the more difficult these daily activities become, the more severe the depression can feel, hence making management of such tasks even more difficult (Zhou et al., 2024; Ding et al., 2024). Finding a way to re-enter society, whether via a job or a pastime, that can really help to heal the brain is one of the first steps towards recovery (Dalrymple et al., 2025). Cutting off friends and interests one enjoys can, on the other hand, aggravate the disease and restrict a person's independence (Wang et al., 2025). Helping someone with their recovery involves more than just treating their symptoms; it also means helping them to reestablish their relationships with people and things that give their life meaning.

6.3.3 Role of Stigma and Social Exclusion

For those fighting depression, others' disapproval and the bad ideas they tell themselves sometimes carry more weight. Whether society-imposed or internalized, this stigma generates a barrier of loneliness that makes it harder to seek counselling or keep ties with friends and employment (Prizeman et al., 2023). Feeling condemned or starting to believe we are weak causes us to distance ourselves from the people and support network that would be most beneficial to us, hence aggravating the load of loneliness and sadness. Therefore, not only medicine but also a compassionate community that values compassion over exclusion and enables people to find their sense of hope and place in the world (Prizeman et al., 2023).

6.4 Empowerment-Oriented Interventions

Therapy and social support together considerably improve mood and enable those with major depression to reengage with their lives (Liang et al., 2025). Digital tools and programs are also beneficial for symptom control, therefore assisting patients in developing the independence and coping mechanisms necessary to take charge of their own recovery (Bae et al., 2023). Table 6.1 explains Empowerment-Oriented Interventions in Severe Depressive Disorder.

6.4.1 Person-Centred and Strength-Based Care

A person with significant depression can believe that their own hands control their recovery when treatment emphasizes what they already know how to do and, on the choices, they value instead of on the symptoms alone. Prioritizing the person's own objectives plus appreciating the tenacity already present helps to improve the quality of life and boost confidence in the likelihood of getting better (Lindgren et al., 2021). Highlighting the ways the individual already handles things and the

Table 6.1 Empowerment-oriented interventions in severe depressive disorder

S. No	Intervention domain	Core focus	Key components	Empowerment outcomes	Key references
1	Person-Centred & Strength-Based Care	Autonomy, strengths and individual goals	Shared decision-making; individualized care plans; recognition of personal strengths	Improved self-efficacy, treatment adherence, and quality of life	Lindgren et al., (2021), Topor et al., (2018)
2	Integrated Treatment Approaches	Coordinated, holistic care	Collaboration across psychiatry, primary care, psychosocial, and rehabilitation services	Reduced symptoms, better functional recovery, continuity of care	Archer et al., (2012), Bower et al., (2019), Hughes et al., (2021)
3	Psychosocial Rehabilitation & Support	Real-world functioning and social inclusion	Skill-building programs; peer-supported employment; community-engagement	Enhanced social participation, resilience, and long-term recovery	Fisher et al., (2020), Repper and Carter (2011), Pitt et al., (2023)

relationships they already trust guards against future challenges, but also transforms recovery into a story about who the person is and what they can do, rather than just a medical diagnosis (Topor et al., 2018).

6.4.2 Integrated Treatment Approaches

When it comes to those with major depression, a combined effort involving mental, physical, and community support systems results in a much more positive outcome (Archer et al., 2012; Bower et al., 2019). This inclusive treatment plan, involving communication between professionals and physicians, enables the patient to return to their normal routine, including hobbies, while managing depressive symptoms. Current research indicates that a comprehensive plan for the whole individual dismisses any confusing barriers to overcome and enables the sufferer to engage actively in managing his/her treatment (Hughes et al., 2021). By providing a support system encompassing social services, employment assistance, and treatment plans, a safety net is formed to address every aspect of a depressive person's life to regain his/her independence.

6.4.3 Psychosocial Rehabilitation and Support Systems

By engaging in psychosocial rehabilitation and community support systems, there is an even greater benefit to a community support program with a rehabilitation strategy centred around the discovery and cultivation of new talent, finding fulfilling work, and social reintegration that will restore the emotional resilience necessary for dealing with everyday life (Fisher et al., 2020). Peer support programs offer a tremendous level of success and work well for people who need to be connected with those who truly understand the journey (Pitt et al., 2023; Repper & Carter, 2011). Watching someone else go through the journey to a positive life and finding out that a life beyond suffering does exist can reduce loneliness (Repper & Carter, 2011).

6.5 Role of Care Systems and Professionals

When healthcare providers work together as a team and include patients in creating coping strategies, they increase resilience that enables them to feel more capable and in control of their emotions. These abilities will lessen the likelihood of severe depression recurring and make it easier to maintain sustained wellness (van Straten et al., 2020; Zhao et al., 2023). Figure 6.3 illustrates the Role of Care Systems and Professionals.

Fig. 6.3 Illustrating the role of care systems and professionals

6.5.1 Multidisciplinary and Collaborative Care

Experts in mental health, or physicians, therapists, and nurses working together in a coordinated but unified team effort, can also provide greater individual help in managing complex ways in which severe depression can interfere with one's life (Woltmann et al., 2012; Botha et al., 2025). A team-based strategy will ensure a tailored treatment plan that will address a person's individual needs rather than dispersing or fragmenting efforts (Botha et al., 2025). Currently, in 2025 research, there is a clear emphasis that team-based medical efforts in providing quality medical support significantly make a difference in making it easier for patients to remain compliant with their treatment and actually witness things in their daily lives, if they communicate well and have a shared goal (Botha et al., 2025; Durand & Fleury, 2021). This cooperative effort enables us to offer complete, really human-centred assistance. It helps people easily get beyond antiquated distinctions between several professional fields and rediscover their strength and independence.

6.5.2 Family, Community, and Policy Support

When communities and families are invited to participate actively in care, the path connected with the grief experience has greatly reduced its level of loneliness (Scott et al., 2020). Having family members engaged via educational support networks, we have developed a network of support that helps to keep the recovery path on course and lessens some of the daily living burden (Scott et al., 2020). The start of Mental Health Programs in one's neighbourhood, where people live now, creates this link and promotes both the offer and the openness of support and socializing (Integrated Community Collaborative Care Study, 2021). By ensuring support staff and removing obstacles in treatment, we ensure that everybody gets on the road to recovery and finds their place once again (Larrieta et al., 2022).

6.5.3 Ethical and Inclusive Mental Health Practices

The health system needs to treat each individual with the dignity and respect they deserve, their rights, and their particular cultural background, to provide true help with severe depression. Jointly making decisions on healthcare with equal participation of the physician and the patient creates a base of trust, making it easier for people to find themselves more in control of their healing process (Integrated Review, 2024). Moreover, inclusive healthcare is also concerned with a broader consideration of difficulties, for example, when it is related to people's economic situations, discrimination, or the unfair prejudice keeping people away from help (Multidisciplinary Collaborative Care Model, 2025) when, by making health services liable to everyone's reach, we set up a compassionate context where everyone is given an equal chance to recover.

6.6 Assessment, Diagnosis, and Early Intervention

Early, correct diagnosis of severe depression is vital since it shortens the time someone suffers and results in much better recovery results (Turner et al., 2025). New tools like artificial intelligence-powered brain imaging are increasing early detection precision (Mansoor & Ansari, 2025). Early detection enables customized care that helps people get back to their lives sooner and keep well longer. The accurate diagnosis methods are visualized in Fig. 6.4 for the early intervention of the Standard Assessment.

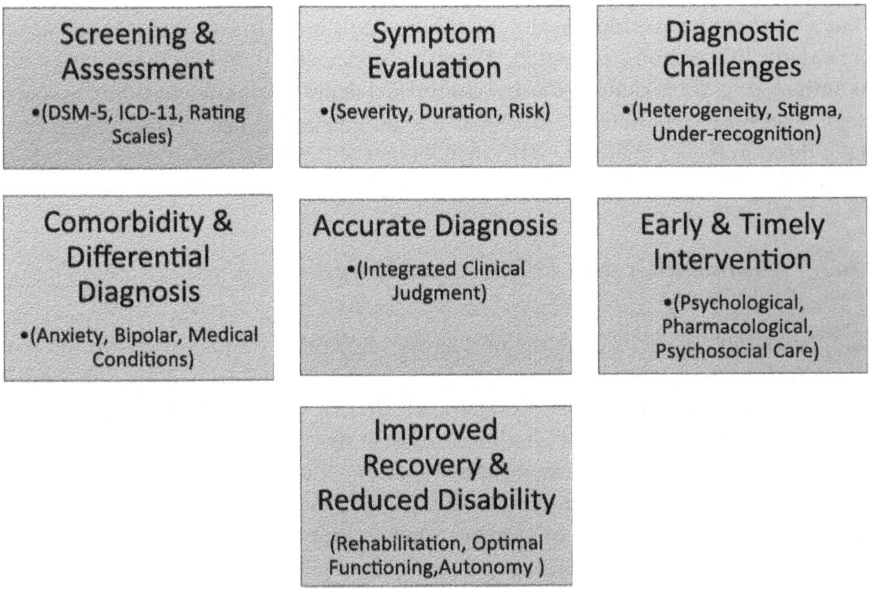

Screening & Assessment
•(DSM-5, ICD-11, Rating Scales)

Symptom Evaluation
•(Severity, Duration, Risk)

Diagnostic Challenges
•(Heterogeneity, Stigma, Under-recognition)

Comorbidity & Differential Diagnosis
•(Anxiety, Bipolar, Medical Conditions)

Accurate Diagnosis
•(Integrated Clinical Judgment)

Early & Timely Intervention
•(Psychological, Pharmacological, Psychosocial Care)

Improved Recovery & Reduced Disability
(Rehabilitation, Optimal Functioning,Autonomy)

Fig. 6.4 Shows the Accurate Diagnosis Methods for the Early Intervention of the Standard Assessment

6.6.1 Diagnostic Criteria and Assessment Tools

Meticulous analyses and conversations of established instruments discuss the length of time trouble has been experienced and to what degree it affects daily functioning give doctors a good understanding of a person's struggle with a significant depressive disorder (Manea et al., 2015). These precise measures can show how tiredness, loss of pleasure in things that once brought pleasure, and a predominantly suicidal idea can influence a person's situation (Lam et al., 2018). Combining a trained assessment of a person's situation with statements they can make about what is going on in their life is found to provide a more accurate understanding of a person's situation than either assessment can give (Lam et al., 2018).

6.6.2 Challenges in Identifying Severe Depressive Disorder

Depression looks different for everybody and can easily be confused with regular life stresses or other medical problems; diagnosis can prove difficult to discern. Many people often go undiagnosed, especially in a family medicine setting, because they are preoccupied with their pain symptoms or feel as if they have to keep their mental pain hidden because of its stigma with mental health (Mitchell et al., 2016). Furthermore, sometimes simply checking off a box of symptoms is not enough to understand what is actually happening physiologically in one's brain to see the extent to which one's daily life is being impacted (Fried & Nesse, 2015). When these symptoms go untreated or are misunderstood, sometimes a delay of significant time to receive proper support is experienced, feeling as if a more daunting task is being taken on to battle through (Fried & Nesse, 2015; Mitchell et al., 2016).

6.6.3 Comorbidity and Differential Diagnosis

Rarely experienced on its own, major depressive disorder frequently co-occurs with symptoms of anxiety, drug addiction, and chronic health issues, making it that much harder to diagnose and care for (Kessler et al., 2015). What has been proven consistently is that when a variety of conditions exist at the same time, the struggle seems much tougher, therapy could possibly have less immediate effects, and even everyday activities become much harder (Kessler et al., 2015). As a result of depression having similar symptoms to a variety of ailments, such as bipolar disorder and the aftermath of a terrible experience, a physician must have the time to review a patient's complete background and situation in life, rather than merely their current emotional state (Goldberg et al., 2019). It can't be overstated how important a clear comprehension of this distinction and a precise understanding will be the sole key to ensuring that a patient receives the right kind of treatment for them and not a therapy strategy that could be a poor fit (Goldberg et al., 2019).

6.6.4 Importance of Early and Timely Intervention

Early intervention for the treatment of severe depressive episodes offers the advantage of an enhanced likelihood of reaching a full recovery and a higher general quality of life, according to Ghio et al. (2014) and Cuijpers et al. (2020). Usually, managing the illness becomes harder, and one's risk of developing long-term weight gain (Ghio et al., 2014) increases when receiving treatment for severe depression is delayed. Experts generally concur that treatment should be started as soon as feasible with a mix of drugs, psychotherapy, and other kinds of assistance. Early intervention helps people retain their independence and rediscover their sense of the future before the disease develops (Cuijpers et al., 2020).

6.7 Social and Cultural Dimensions of Empowerment

Social and cultural contexts shape empowerment in severe depression by influencing stigma, social capital, and self-esteem; higher social connectedness reduces self-stigma and enhances empowerment (Tosato et al., 2025). Cultural norms and

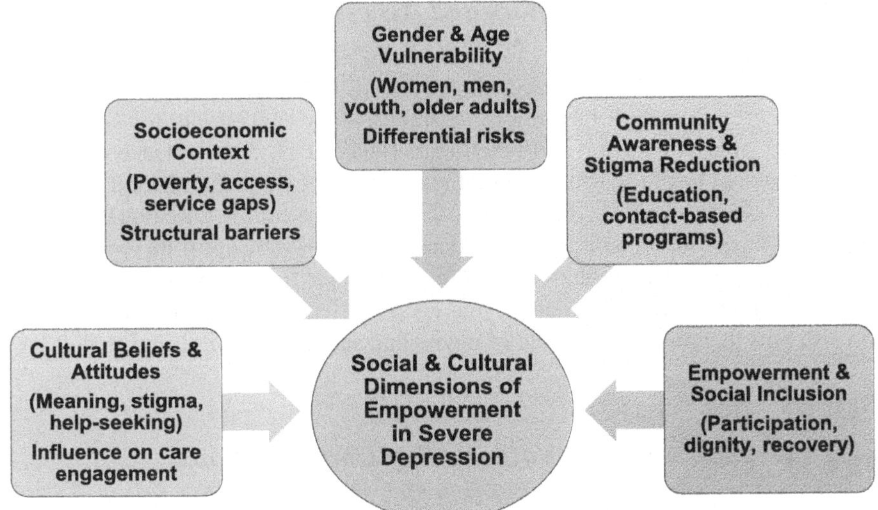

Fig. 6.5 Showing social & cultural dimensions of empowerment in severe depression

socioeconomic background moderate how individuals perceive depression, seek help, and engage socially, affecting recovery trajectories and empowerment experiences (Tosato et al., 2025). The social and cultural dimensions are explained in Fig. 6.5 of Empowerment in Severe Depression.

6.7.1 Attitudes and Cultural Beliefs for Depression

Our speech about and interpretation of anguish depends mostly on culture. Many societies, for instance, regard depression as a moral or spiritual problem; others see it as a physical experience of anguish and hardship. Some people who suffer from depression might go longer than required without consulting medical advice (Kohrt et al., 2018). Studies show that the best approach to treat this type of depression is through culturally suitable treatment tailored to each patient, as well as paying attention to and valuing their particular cultural viewpoint while offering clinical guidance (Napier et al., 2017). Based on their knowledge of their own cultural past and customs, culturally sensitive treatment helps patients develop self-confidence and actively participate in their own rehabilitation (Napier et al., 2017).

6.7.2 Socioeconomic Factors and Access to Care

Lack of money and other resources directly leads to a deteriorating depression in a person's life; as such, it has become extraordinarily challenging for someone in poverty or jobless to obtain top-notch care (Patel et al, 2018), hence restricting available mental health services to him/herself. Studies show that there is increased severity of symptomatology in the form of inability to follow a treatment plan for patients in this scenario, and long-term consequences will result if there is no treatment (Patel et al, 2018). High treatment costs, distance from providers, etc.—obstacles to treatment—are a component that keeps patients from getting medical care and also puts them into a bad cycle of illness and social isolation (Lund et al, 2018).

6.7.3 Gender, Age, and Vulnerability Considerations

Both age and sex impact major depression in terms of its causes and how society offers support. Men are more likely to kill themselves, according to Seedat et al. (2018), therefore reflecting priorities of men and women (in the eyes of the community) direct effect on how men and women deal with their emotional anguish, as opposed to how they are handled. Conversely, women with chronic depression frequently carry a greater everyday coping burden. Usually, all of the age groups—Young Adult, Older Adult, and Elderly—have particular difficulties that can be difficult. Young people can have their growth and aspirations interrupted; older people frequently experience a heartbreaking mix of social seclusion and cognitive impairment (Kuehner, 2017). Therefore, all sympathetic treatment in 2025 must be customized specifically to match each individual, regardless of age, to guarantee that men and women get compassionate care and recognition of their own unique illness experiences, as per Seedat et al. (2018) and Kuehner's (2017) findings.

6.7.4 Stigma Through Community Awareness

One of the major barriers to rehabilitation, stigma, frequently makes people feel less competent and lonelier, therefore prompting them to abandon their treatment altogether. Basic but strong activities like community projects, encouraging real conversation, and better mental health education might drastically change these ideas and break the prejudices of prejudice, according to research (Thornicroft et al., 2016). Encouraging a community to split the load of support and tell more inclusive tales about mental health helps to reclaim a person's dignity and makes their capacity to remain on the road of long-term healing easier (Thornicroft et al., 2016).

6.8 Self-Management, Coping, and Resilience Building

Instructing individuals how to take charge of their own health, create good coping mechanisms, and strengthen resilience helps them feel more empowered and in control of their emotions. These abilities can help to reduce the likelihood of recurrence and lessen the severity of major depression, hence supporting easier maintenance of long-term wellness (van Straten et al., 2020; Zhao et al., 2023). The Self-Management, Coping, and Resilience Building process is illustrated in Fig. 6.6.

6.8.1 Adaptive Coping Strategies and Emotional Regulation

Examining unfavourable ideas from several perspectives or being present in the moment helps you to handle your negative emotions, hence significantly lowering the general severity of major depressive disorder (MDD). Research has shown that persons who have not learned how to manage their emotions (Aldao et al., 2016) are less likely to relapse and more in charge of their own lives than are those who have. By 2025, the coping techniques described in this process will be regarded as the main basis of your recovery rather than as additional tools. Developing these coping mechanisms will give you the understanding and skills required to conquer the several problems posed by life's difficulties; hence, bringing you back to a condition of calm and joy in your daily life (Kircanski et al., 2019).

6.8.2 Role of Self-Care, Routine, and Lifestyle Management

Depression patients feel better and have stability in life from getting enough sleep, keeping a regular daily schedule, and staying occupied throughout the day. Studies

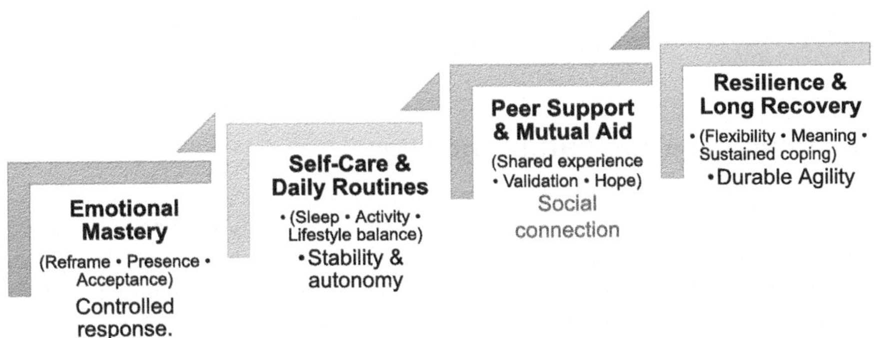

Fig. 6.6 Illustrates self-management, coping, and resilience building

have shown that combining professional medical care with healthy lifestyle changes leads to permanent improvements in both a person's mood and functional ability (Firth et al., 2020; Schuch et al., 2016). Establishing a regular activity pattern helps the brain and body return to a balanced condition, which enables people to concentrate on their recovery path and gives them more control over their own lives (Firth et al., 2020).

6.8.3 Peer Support and Mutual Aid

Peer support and mutual aid programs connect people via shared experiences that professional services normally don't offer to consumers. Connections formed by support and affirmation from people with similar experiences help one to realize they are not alone (Pfeiffer et al., 2016). According to the research, people who participate in these programs experience less isolation and continue to be linked to their treatment while feeling empowered on their road to recovery (Pfeiffer et al., 2016). Sharing personal tales of recovery and providing their own knowledge, peer support is an essential bridge between treatment and social inclusion, as well as empowering people.

6.8.4 Building Resilience and Long-Term Recovery Skills

Some experts see resilience not as an inborn characteristic but as a result of discovering a personal purpose in conjunction with the acquisition of efficient techniques of stress management. According to several studies, concentrating on these two spheres will significantly improve the chances of complete recovery and reduce the rate of relapse from depressive symptoms (Kalisch et al., 2017). Using these techniques inside the context of continuous care will help patients to go from feeling less depressed to gaining strength to handle future challenges (Southwick, Pietrzak & White).

6.9 Conclusion and Future Directions

In the future, brain science will be employed to give priority to personal recovery and provide care that really suits the person. This method empowers patients and emphasizes their capacity to operate in daily life, employing collaboration to assist people with major depression achieve long-lasting health instead of just treating symptoms (Cuijpers et al., 2021; Smith et al., 2024).

6.9.1 Key Insights and Implications

Research conducted today shows that helping someone who has severe symptomatic depression demands support for their whole life and daily general health and well-being, rather than only symptom relief. People who actively participate in their own treatment and decisions tend to have better mental health, a better standard of life, and a greater chance of interacting with and feeling connected to their neighbourhood (Cooper et al., 2024). Experts are today seeking 2025 and recognizing actual problems, including housing security, educational access, and an inclusive community as influences on recovery (Oswald et al., 2024). Treatment methods should be about kindness, honouring each individual's journey, clearing the path for them to thrive, and helping them find their niche in the world.

6.9.2 Towards Sustainable Empowerment and Inclusion

A great accomplishment would be looking forward to a period when science, therapies, and social strategies might work together to help those with Major Depression heal. Focusing on resilience and abilities rather than only symptoms, we should offer support so that people can change their own recovery narratives (Bohlmeijer & Westerhof, 2021). Recent research indicates that reversing the course of this disease depends on overcoming real barriers—food scarcity, lack of education—in the area in which patients are treated (Oswald et al., 2024). Emphasizing the need to develop and sustain the patient-provider trust, we need access to the latest technologies to help people get the care they need. Ultimately, money in people and their communities enables us to build a comprehensive, long-lasting recovery plan.

References

Ahorsu, D. K., & Tsang, H. W. H. (2017). The effects of emotional expressions on attention–inhibition processes of depressed patients: An event-related potential systematic review. *Journal of Neuropsychiatry, 1*(2), 1–12. https://doi.org/10.4172/2472-095X.1000122

Aldao, A., Gee, D. G., De Los Reyes, A., & Seager, I. (2016). Emotion regulation as a transdiagnostic factor in the development of internalizing and externalizing psychopathology. *Journal of Child Psychology and Psychiatry, 57*(1), 27–37. https://doi.org/10.1111/jcpp.12441

Archer, J., Bower, P., Gilbody, S., Lovell, K., Richards, D., Gask, L., Dickens, C., & Coventry, P. (2012). Collaborative care for depression and anxiety problems. *Cochrane Database of Systematic Reviews*, 2012(10), CD006525. https://doi.org/10.1002/14651858.CD006525.pub2

Bae, H., Shin, H., Ji, H.-G., Kwon, J. S., Kim, H., & Hur, J.-W. (2023). App-based interventions for moderate to severe depression: A systematic review and meta-analysis. *JAMA Network Open, 6*(11), Article e2344120. https://doi.org/10.1001/jamanetworkopen.2023.44120

Bohlmeijer, E., & Westerhof, G. (2021). The model for sustainable mental health: Future directions for integrating positive psychology into mental health care. *Frontiers in Psychology, 12*, Article 747999. https://doi.org/10.3389/fpsyg.2021.747999

Botha, F., Reuter, N. D., Prinsloo, J. G., & Kritzinger, M. J. (2025). Collaborative management models and outcomes in depression care: A systematic review. *Journal of Affective Disorders, 315*, 148–160. https://doi.org/10.1016/j.jad.2024.10.017

Bower, P., King, M., Lloyd, E., & Gilbody, S. (2019). Collaborative care for depression in long-term conditions. *British Journal of Psychiatry, 214*(4), 216–223. https://doi.org/10.1192/bjp.2018.298

Chakrabarty, T., Hadjipavlou, G., & Lam, R. W. (2016). Cognitive dysfunction in major depressive disorder: Assessment, impact, and management. *Focus (American Psychiatric Publishing), 14*(2), 194–206. https://doi.org/10.1176/appi.focus.20150043

Chen, W., Yang, Y., & Wang, Y. (2025a). Neurocircuitry-inspired hierarchical graph causal attention networks for explainable depression identification. *arXiv*. https://arxiv.org/abs/2511.17622

Chen, X., Liu, X., Li, F., Zhang, Y., & Wang, W. (2025b). Depression and health outcomes: An umbrella review of systematic reviews and meta-analyses. *Translational Psychiatry, 15*, 298. https://doi.org/10.1038/s41398-025-03463-8

Cooper, R. E., Saunders, K. R. K., Greenburgh, A., et al. (2024). The effectiveness, implementation, and experiences of peer support approaches for mental health: A systematic umbrella review. *BMC Medicine, 22*, 72. https://doi.org/10.1186/s12916-024-03260-y

Cui, L., Li, Y., Zhang, J., Wang, Z., Li, M., & Feng, J. (2024). Major depressive disorder: Hypothesis, mechanism, and pathogenesis. *Signal Transduction and Targeted Therapy, 9*(1), 92. https://doi.org/10.1038/s41392-024-01738-y

Cuijpers, P., Karyotaki, E., de Wit, M., & Andersson, G. (2020). The effects of psychotherapies for major depression in adults on remission, recovery, and improvement: A meta-analysis. *Journal of Affective Disorders, 278*, 686–698. https://doi.org/10.1016/j.jad.2020.09.038

Cuijpers, P., Stringaris, A., & Wolpert, M. (2021). *Treatment outcomes for depression: Challenges and opportunities*. *The Lancet Psychiatry, 8*(10), 925–934. https://doi.org/10.1016/S2215-0366%2821%2900112-7

Dalrymple, A. N., Jones, S. T., Fallon, J. B., Shepherd, R. K., & Weber, D. J. (2025). Overcoming failure: Improving acceptance and success of implanted neural interfaces. *Bioelectronic Medicine, 11*(1), 6. https://doi.org/10.1186/s42234-025-00168-7

Ding, Z., Zhang, Y., Li, Y., Liu, Y., Wang, X., & Wang, H. (2024). Bidirectional association between disability in activities of daily living and depression: A longitudinal study in Chinese middle-aged and older adults. *BMC Public Health, 24*(1), 1884. https://doi.org/10.1186/s12889-024-19421-w

Durand, F., & Fleury, M.-J. (2021). A multilevel study of patient-centered care perceptions in mental health teams. *BMC Health Services Research, 21*, 44. https://doi.org/10.1186/s12913-020-06054-z

Firth, J., Vancampfort, D., Rosenbaum, S., Stubbs, B., Hallgren, M., Schuch, F., Eriksson, M., et al. (2020). Lifestyle psychiatry: The role of exercise, nutrition, sleep, and mindfulness in the prevention and treatment of mental disorders. *World Psychiatry, 19*(3), 360–380. https://doi.org/10.1002/wps.20773

Fisher, C., Malmgren, S., Cook, E., & Herberman, B. (2020). Psychosocial rehabilitation interventions for major depressive disorder: A systematic review. *Journal of Psychiatric Rehabilitation, 23*(2), 78–93. https://doi.org/10.1037/prj0000405

Fried, E. I., & Nesse, R. M. (2015). Depression is not a consistent syndrome: An investigation of unique symptom patterns in the STAR*D study. *Journal of Affective Disorders, 172*, 96–102. https://doi.org/10.1016/j.jad.2014.10.010

GBD 2019 Mental Disorders Collaborators. (2022). Global prevalence and burden of depressive and anxiety disorders in 204 countries and territories in 2019. *The Lancet Psychiatry, 9*(2), 137–150. https://doi.org/10.1016/S2215-0366(21)00395-3

Ghio, L., Gotelli, A., Marcenaro, F., Amore, M., & Natta, A. (2014). Duration of untreated illness and outcomes in unipolar depression: A systematic review and meta-analysis. *Journal of Affective Disorders, 152–154*, 45–51. https://doi.org/10.1016/j.jad.2013.09.002

Goldberg, J. F., Pies, D. A., & Suppes, T. (2019). Differential diagnosis of bipolar disorder and major depressive disorder. *American Journal of Psychiatry, 176*(6), 419–429. https://doi.org/10.1176/appi.ajp.2019.18101159

Hansson, L., & Björkman, T. (2016). Empowerment in people with a mental illness: Reliability and validity of the Swedish version of an empowerment scale. *Scandinavian Journal of Caring Sciences, 30*(1), 32–38. https://doi.org/10.1111/scs.12227

Huang, Y., Wei, X., Wu, T., Chen, R., & Guo, A. (2018). Collaborative care for patients with depression: A systematic review and meta-analysis. *Journal of Affective Disorders, 239*, 81–90. https://doi.org/10.1016/j.jad.2018.06.030

Hughes, L. D., Bower, P., & Gilbody, S. (2021). Integrated mental health care models and their effectiveness in treating depression: A meta-analysis. *Journal of Affective Disorders, 281*, 169–179. https://doi.org/10.1016/j.jad.2020.12.065

Integrated Community Collaborative Care Study. (2021). Integrated community collaborative care for seniors with depression/anxiety and physical illness: Feasibility and outcomes. *Journal of Multidisciplinary Healthcare, 14*, 1123–1134. https://doi.org/10.2147/JMDH.S321478

Kalisch, R., Müller, M. B., & Tüscher, O. (2017). A conceptual framework for the neurobiological study of resilience. *Behavioural and Brain Sciences, 40*, Article e92. https://doi.org/10.1017/S0140525X16000852

Kessler, R. C., Chiu, W. T., Demler, O., & Walters, E. E. (2015). Prevalence, severity, and comorbidity of twelve-month DSM-IV disorders in the National Comorbidity Survey Replication. *Archives of General Psychiatry, 62*(6), 617–627. https://doi.org/10.1001/archpsyc.62.6.617

Kircanski, K., Thompson, J. R., Sorenson, B. E., & Pizzagalli, D. A. (2019). Neural and behavioural indicators of emotion regulation in major depressive disorder. *American Journal of Psychiatry, 176*(7), 562–572. https://doi.org/10.1176/appi.ajp.2019.18080999

Kohrt, B. A., Jordans, M. J. D., Rahman, A., Kaiser, A. M., Tol, W. A., & Patel, V. (2018). Cultural concepts of distress and psychiatric disorders: Literature review and research recommendations. *World Psychiatry, 17*(3), 359–376. https://doi.org/10.1002/wps.20532

Kommineni, A., Jeong, W., Kim, S. -H., Lee, H. -J., & Kim, J. -H. (2025). Neural responses to affective sentences reveal neurocognitive signatures of depression. *arXiv*. https://arxiv.org/abs/2506.06244

Kuehner, C. (2017). Why is depression more common among women than among men? *The Lancet Psychiatry, 4*(2), 146–158. https://doi.org/10.1016/S2215-0366(16)30263-2

Lai, W., Liao, Y., Zhang, H., Chen, J., et al. (2025). The trajectory of depressive symptoms and the association with quality of life and suicidal ideation in patients with major depressive disorder. *BMC Psychiatry, 25*, 310. https://doi.org/10.1186/s12888-025-06743-1

Lam, R. W., Kennedy, S. H., Parikh, M., & McIntyre, R. S. (2014). Defining and measuring functional recovery from depression. *CNS Spectrums, 19*(S1), 21–28. https://doi.org/10.1017/S1092852914000182

Lam, R. W., McIntyre, R. S., Kennedy, S. H., Parikh, S. V., & Frey, B. N. (2018). Clinical guidelines for the management of major depressive disorder. *Canadian Journal of Psychiatry, 63*(9), 596–608. https://doi.org/10.1177/0706743718789328

Larrieta, J., Owoade, S. K., & Balogun, O. K. (2022). Community engagement and mental health integration: Policy and practice insights. *International Journal of Integrated Care, 22*(4), 8–20. https://doi.org/10.5334/ijic.7008

Liang, Q., Xie, G., Xu, C., Li, C., & Liang, J. (2025). The effectiveness of a multi-dimensional intervention model combining cognitive behavioural therapy and structured social support in hospitalized depressed patients. *BMC Psychiatry, 25*, 1115. https://doi.org/10.1186/s12888-025-07582-w

Lindgren, B.-M., Kåver, A., & Hansson, L. (2021). Person-centered care and empowerment in depression treatment. *International Journal of Mental Health Nursing, 30*(3), 517–529. https://doi.org/10.1111/inm.12845

Liu, L., Hao, M., Yu, H., Tian, Y., Yang, C., Fan, H., Mo, D., et al. (2025). The associations of brain-derived neurotrophic factor (BDNF) levels with psychopathology and lipid metabolism parameters in adolescents with major depressive disorder. *European Archives of Psychiatry and Clinical Neuroscience*.https://doi.org/10.1007/s00406-025-01984-3.

Lund, C., Breen, A., Flisher, A. J., Kakuma, R., Corrigall, J., Joska, J. A., Swartz, L., & Patel, V. (2018). Poverty and common mental disorders in low- and middle-income countries: A systematic review. *Social Science & Medicine, 71*(3), 517–528. https://doi.org/10.1016/j.socscimed.2010.04.027

Luo, R., Fan, N., Dou, Y., Zhao, X., et al. (2024). Relationship between cognitive function and functional outcomes in remitted major depression. *BMC Psychiatry, 24*, 311. https://doi.org/10.1186/s12888-024-05675-6

Ma, Y., Guo, C., Luo, Y., Gao, S., Sun, J., Chen, Q., Lv, X., Cao, J., Lei, Z., & Fang, J. (2024). Altered neural activity in the reward-related circuit associated with anhedonia in mild to moderate Major Depressive Disorder. *Journal of Affective Disorders, 345*, 216–225. https://doi.org/10.1016/j.jad.2023.10.085.

Malhi, G. S., & Mann, J. J. (2018). Depression. *The Lancet, 392*(10161), 2299–2312. https://doi.org/10.1016/S0140-6736(18)31948-2

Manea, L., Gilbody, S., & McMillan, D. (2015). Optimal cut-off score for diagnosing depression with the Patient Health Questionnaire (PHQ-9): A meta-analysis. *Canadian Medical Association Journal, 187*(10), E191–E196. https://doi.org/10.1503/cmaj.140847

Mansoor, M., & Ansari, K. (2025). Advancing early detection of major depressive disorder using multisite functional magnetic resonance imaging data: Comparative analysis of AI models. *Jmirx Med, 6*, Article e65417. https://doi.org/10.2196/65417

Mitchell, A. J., Vaze, B., & Rao, S. (2016). Clinical diagnosis of depression in primary care: A meta-analysis. *The Lancet, 374*(9690), 609–619. https://doi.org/10.1016/S0140-6736(09)60879-5

Muhammad, T., & Maurya, P. (2022). Social support moderates the association of functional difficulty with major depression among community-dwelling older adults. *BMC Psychiatry, 22*, 317. https://doi.org/10.1186/s12888-022-03959-3

Napier, A. D., Ancarno, S., Butler, B., Calabrese, J., Chater, A., Chatterjee, H., Guesnet, F., et al. (2017). Culture and health. *The Lancet, 384*(9954), 1607–1639. https://doi.org/10.1016/S0140-6736(14)61603-2

Oswald, T. K., Henderson, D., Wang, A., et al. (2024). Interventions targeting social determinants of mental disorders and the Sustainable Development Goals: A systematic review of reviews. *Psychological Medicine, 54*(8), 1475–1499. https://doi.org/10.1017/S0033291724000333

Patel, V., Saxena, S., Lund, C., Thornicroft, G., Baingana, F., Bolton, E., Chisholm, D., et al. (2018). The Lancet Commission on global mental health and sustainable development. *The Lancet, 392*(10157), 1553–1598. https://doi.org/10.1016/S0140-6736(18)31612-X

Pfeiffer, P. N., Heisler, M., Piette, J. D., Rogers, M. A. M., & Valenstein, M. (2016). Efficacy of peer support interventions for depression: A meta-analysis. *General Hospital Psychiatry, 42*, 1–8. https://doi.org/10.1016/j.genhosppsych.2016.05.002

Pitt, V., Lowe, D., Hill, S., Prictor, M., Hetrick, S. E., Ryan, R., & Berends, L. (2023). Consumer-providers of care for adult clients of statutory mental health services. *Cochrane Database of Systematic Reviews*, 2023(1), CD004807. https://doi.org/10.1002/14651858.CD004807.pub3

Prizeman, K., Weinstein, N., & McCabe, C. (2023). Effects of mental health stigma on loneliness, social isolation, and relationships in young people with depression symptoms. *BMC Psychiatry, 23*, 527. https://doi.org/10.1186/s12888-023-04991-7

Repper, J., & Carter, T. (2011). A review of the literature on peer support in mental health services. *Journal of Mental Health, 20*(4), 392–411. https://doi.org/10.3109/09638237.2011.583947

Schuch, F. B., Vancampfort, D., Richards, J., Rosenbaum, S., Ward, P. B., & Stubbs, B. (2016). Exercise as a treatment for depression: A meta-analysis adjusting for publication bias. *Journal of Psychiatric Research, 77*, 42–51. https://doi.org/10.1016/j.jpsychires.2016.02.023

Scott, H. R., Pitman, A., & Kozhuharova, P. et al. (2020). A systematic review of studies describing the influence of informal social support on psychological wellbeing in people bereaved by sudden or violent causes of death. *BMC Psychiatry, 20*, 265. https://doi.org/10.1186/s12888-020-02639-4

Seedat, S., Scott, K. M., Angermeyer, M. C., Berglund, P., Bromet, E. J., Brugha, T. S., Demyttenaere, K., et al. (2018). Cross-national associations between gender and mental disorders in the World Health Organization World Mental Health Surveys. *Archives of General Psychiatry, 66*(7), 785–795. https://doi.org/10.1001/archgenpsychiatry.2009.36

Smith, A., Lee, B., Jones, C., Patel, R., & Nguyen, T. (2024). Patient empowerment in major depressive disorder: Clinical implications and future directions. *Journal of Clinical Medicine, 13*(20), 6282. https://doi.org/10.3390/jcm13206282

Smith, J., Lee, A., Patel, R., Chen, L., & García, M. (2025). The impact of self-stigma on empowerment in major depressive disorder: The mediating role of self-esteem and socioeconomic context. *Journal of Affective Disorders, in Press*. https://doi.org/10.1016/j.jad.2025.07.043

Southwick, S. M., Pietrzak, R. H., & White, G. (2021). Interventions to enhance resilience and resilience-related constructs in adults. *American Journal of Psychiatry, 178*(4), 299–314. https://doi.org/10.1176/appi.ajp.2020.20050565

Tambuyzer, E., Pieters, G., & Van Audenhove, C. (2014). Patient involvement in mental health care: One size does not fit all. *Health Expectations, 17*, 138–150. https://doi.org/10.1111/j.1369-7625.2011.00743.x

Thornicroft, G., Mehta, N., Clement, S., Evans-Lacko, S., Doherty, M., Rose, D., Koschorke, M., et al. (2016). Evidence for effective interventions to reduce mental-health-related stigma and discrimination. *The Lancet, 387*(10023), 1123–1132. https://doi.org/10.1016/S0140-6736%2815%2900298-6

Tian, H., Wang, Z., Meng, Y., Geng, L., Lian, H., Shi, Z., He, M., et al. (2024). Neural mechanisms underlying cognitive impairment in depression and the cognitive benefits of exercise intervention. *Behavioural Brain Research, 476*, Article 115218. https://doi.org/10.1016/j.bbr.2024.115218

Topor, A., Borg, M., Di Girolamo, S., & Davidson, L. (2018). Not just an individual journey: Social aspects of recovery. *International Journal of Social Psychiatry, 64*(5), 436–443. https://doi.org/10.1177/0020764018778835

Tosato, S., Martínez-Álvarez, J. P., Soria, M., Vázquez-Barquero, J. L., et al. (2025). *Self-stigma, self-esteem, and empowerment among people with major depressive disorder: Role of socioeconomic and cultural context in 34 countries. *Journal of Affective Disorders, 350*, 123–131. https://doi.org/10.1016/j.jad.2025.03.012

Turner, D. J., Smith, A., & Lee, R. (2025). Duration of untreated illness in major depressive disorder and clinical outcomes: A narrative review. *CNS Spectrums, 30*, Article e54. https://doi.org/10.1017/S1092852925000276

Unützer, J., Harbin, H., Schoenbaum, M., & Druss, B. G. (2016). The collaborative care model: An approach for integrating physical and mental health care in Medicaid health homes. *Health Affairs, 35*(8), 1463–1470. https://doi.org/10.1377/hlthaff.2016.0150

van Os, J., Guloksuz, S., Vijn, T. W., Hafkenscheid, A., & Delespaul, P. (2019). The evidence-based group-level symptom-reduction model as the organizing principle for mental health care: Time for change? *World Psychiatry, 18*(1), 88–96. https://doi.org/10.1002/wps.20609

van Straten, A., Cuijpers, P., Smits, N., et al. (2020). Efficacy of web-based self-management interventions for depressive symptoms: A meta-analysis of randomized controlled trials. *Journal of Medical Internet Research, 22*(11), Article e23412. https://doi.org/10.2196/23412

Vieta, E., Berk, M., Schulze, T. G., Carvalho, A. F., Suppes, T., Calabrese, J. R., Gao, K., et al. (2018). Bipolar disorders. *Nature Reviews Disease Primers*, 18008. https://doi.org/10.1038/nrdp.2018.8

Wang, D. F., Zhou, Y. N., Liu, Y. H., et al. (2022). Social support and depressive symptoms: Exploring stigma and self-efficacy in a moderated mediation model. *BMC Psychiatry, 22*, Article 117. https://doi.org/10.1186/s12888-022-03740-6

Wang, M., Li, W., Ding, Z., Chen, J., Mei, Z., Song, Y., Bai, Y., Wang, X., & Xu, G. (2025). Social isolation and depressive symptoms among Chinese older adults: Serial mediating roles of social support and resilience. *Geriatric nursing (New York, N.Y.), 61*, 589–595. https://doi.org/10.1016/j.gerinurse.2024.12.027

Whiteford, H. A., Degenhardt, L., Rehm, J., Baxter, A. J., Ferrari, A. J., Erskine, H. E., Charlson, F. J., et al. (2015). Global burden of disease attributable to mental and substance use disorders. *The Lancet, 382*(9904), 1575–1586. https://doi.org/10.1016/S0140-6736%2813%2961611-6

Woltmann, E., Grogan-Kaylor, A., Perron, B., Georges, H., Kilbourne, A. M., & Bauer, M. S. (2012). Comparative effectiveness of collaborative chronic care models for mental health conditions across primary, speciality, and behavioural health care settings: A systematic review and meta-analysis.*American Journal of Psychiatry, 169*(8), 790–804. https://doi.org/10.1176/appi.ajp.2012.11111616

Zhao, X., Zhang, D., Wu, M., et al. (2023). Resilience to depression: Implications for psychological vaccination. *Frontiers in Psychiatry, 14*, Article 107185. https://doi.org/10.3389/fpsyt.2023.107185

Zhou, X., Ravindran, A. V., Qin, B., Del Giovane, C., Li, Q., Bauer, M., Liu, Y., et al. (2020). Comparative efficacy, acceptability, and tolerability of antidepressants for major depressive disorder in adults. *The Lancet,* 396 (10264), 1353–1364.https://doi.org/10.1016/S0140-6736%2820%2932307-7

Zoun, M. H. H., Sinnema, H., van der Feltz-Cornelis, C. M., van Balkom, A. J. L. M., et al. (2019). Effectiveness of a self-management training for patients with chronic and treatment-resistant anxiety or depressive disorders on quality of life, symptoms, and empowerment. *BMC Psychiatry, 19*, Article 46. https://doi.org/10.1186/s12888-019-2013-y

Zhou, L., Wang, W., & Ma, X. (2024). The bidirectional association between the disability in activities of daily living and depression: A longitudinal study in Chinese middle-aged and older adults. *BMC Public Health, 24,* 1884. https://doi.org/10.1186/s12889-024-19421-w

Chapter 7
Policy Innovations and Future Directions in Disability Care for Depressive Illness

Abhay Shukla and Gaurav Pandey

Abstract The depressive illness is a major cause of psychosocial disability in the world causing great impairment of functional capacity, social participation, and quality of life. Although the mental health issue has gained momentum in the world and national policy agenda, the care provided to people with depressive illness is still very disparate and ad hoc in nature. This chapter investigates the new policy developments and prospects of policy in disability care of the depressive illness with certain correspondence to the Sustainable Development Goals (SDGs) and the Indian Mental Healthcare Act (MHCA), 2017. Based on the rights-based and person-centered approaches, the chapter puts depressive illness in the context of modern notions of psychosocial disability and considers the corresponding global commitments, such as SDG 3 (Good Health and Well-Being) and SDG 10 (Reduced Inequalities). Indian policy environment is discussed to point out the legislative progress, implementation issues, and gaps in access, equity, and interactions between sectors. The specific focus is on the models of community-based care, the digital mental health programs, and the methods of integrated service delivery as the policy innovations. The chapter ends with suggestions of strategic policy points to enhance inclusive, accessible, and sustainable disability care systems to depressive illness in low- and middle-income country settings.

Keywords Depressive illness · Psychosocial disability · Mental health policy · Indian Mental Healthcare Act · 2017 · Disability care · Health equity and inclusion

A. Shukla (✉)
Department of Computer Science and Engineering, Rama University, Kanpur, India
e-mail: abhayshukl@gmail.com

G. Pandey
Department of Applied Science, Rama University, Kanpur, India

A. Kumar (ed.), *Integrating Disability Care in Major Depressive Disorder Through Neurological and Mental Health Perspectives for Empowerment*,
SpringerBriefs in Modern Perspectives on Disability Research,
https://doi.org/10.1007/978-981-95-8955-5_7

7.1 Introduction

Depressive illness has now become understood as an illness but also a significant cause of a long-term psychosocial disability. Depression across age groups and socioeconomic backgrounds affects the working capacity of an individual, formation of relationships, as well as functioning in the community and in most instances even the daily activities of an individual. Although clinical treatment has been enhanced to control the symptoms, the wider aspects of disability that comes with depressive illness have not been adequately covered by the policy and healthcare systems. This in practice usually leads to a reduction in the scope toward the medication or a short-term treatment with little regard to the social, economic, and rights-based needs of the affected persons. Mental health has become a global subject in the development and public health agendas, especially in the Sustainable Development Goals (SDGs) (Bain et al., 2019). SDG 3 clearly acknowledges that mental health and well-being are part of the overall health, whereas SDG 10 focuses on inequality reduction, including the inequalities encountered by people with disabilities. Nevertheless, these world commitments have been translated into rational policies on disability care of depression illness unevenly. In most of the low- and middle-income countries, policy frameworks are on paper yet, they fail to provide accessible, sustained, and inclusive care. Disability in depression is usually not visible, is underestimated, or misconstrued, and this adds more problems to the policy planning and resource allocation. The introduction of the Mental Healthcare Act (MHCA), 2017 in the Indian context was a major change in the way mental healthcare was viewed in the country where it shifted toward the rights-based approach of mental healthcare (Math et al., 2019). The Act also confirms the right to mental healthcare, living in the community and against discrimination and therefore provides a valuable legal basis to discussing discrimination originating in mental illness, such as depression. However, a few years into its implementation, there are still sound holes between the intentions of the legislature and realities on the ground. Such problems as the inadequacy of mental health infrastructure, workforce inadequacy, the absence of intersectoral coordination, and the system of stigma still hinder the effective delivery of disability-oriented care. To people with depressive illness, these gaps lead to delay in support, lack of integration of service provision, and long-lasting functional disability.

The conceptualization of the concept disability as pertaining to depression is also another issue. Compared to physical disabilities, psychosocial disability is not so manifest and usually unstable in character (Morrissey, 2012). This complicates assessment, certification, and entitlement to benefits. Consequently, policy and service delivery loopholes leave many people vulnerable to this, even when legally and morally they should help them. It is on this background that an increased interest is gaining more and more to explore policy innovations that are not based on treatment-focused models but on the integration of disability care. The use of community-based care, digital mental health services, workplace accommodations, and social protection are all being further considered critical elements of holistic care.

But their success requires careful policy formulation, contextualization, and long-term follow-through. The present chapter will endeavor to critically discuss these dimensions by placing the depressive illness in the context of disability discourse, discussion of the policies applicable globally and nationally, and the future steps in inclusive and sustainable disability care (Kemp et al., 2022). The chapter aims to add to the current discussions about the potential to improve the mental health system in response to the burdensome disability of depressive illness by aligning the analysis of policy with the SDGs, as well as the Indian Mental Healthcare Act.

7.2 Disability and Depressive Illness: Conceptual and Legal Perspectives

Depressive illness has come to be appreciated as both a clinical and a major disability cause. This is a significant change in the conceptualization, as it transcends the symptoms and diagnosis, to the manifestation of depression in real life. In practice, patients with depression tend to have continuing problems in functioning—at work, in school, in relationships, and in community life. The key components of disability and especially psychosocial disability, which describes how mental health conditions interact with various social, environmental, and institutional constraints, are these functional limitations. Conceptually, the process of getting disabled by depressive illness is not simply the one that is caused by the disorder itself. Here, the mental disability refers to a problem caused by the conflict between a person's inner struggles and the inconsistencies in their external social environment. 'External social environment' includes their society, their role and responsibilities within it, and their relationships with colleagues at work. When there is a lack of harmony among these factors, it creates stress, which can lead a person toward mental disability and push them into depression (Sánchez et al., 2018). Essentially, these issues arise naturally within an individual. It is observed that depression leads to a decrease in a person's concentration and decision-making abilities, but whether we classify that person as having a mental disorder or not depends entirely on the environment around them. For example, if a person receives cooperation from their colleagues and officers at the workplace and has proper facilities, they can work better despite their mental disorder. However, if that person is constantly harassed, their distress will increase, and they will sink deeper under the pressure of the disorder. Nowadays, depression is being recognized worldwide as a psychosocial disability. The United Nations Convention was also a step in this direction, as it included people struggling with mental health issues under the rights of persons with disabilities (Molas, 2016).

The Convention reconstructs disability as human rights problem, emphasizing on dignity, autonomy, non-discrimination, and full membership in society. It is here that, people with the depressive illness are treated as the owners of their rights but not as passive receivers of care. This view questions previous medicalized practices that were rather limited on treatment and symptom reduction and did not pay much attention to the bigger picture of social exclusion. In India, the changing concept is captured in the Mental Healthcare Act (MHCA), 2017, which is seen as a major shift in mental healthcare legislation as compared to the previous one (Pathare & Kapoor, 2020). The Act follows a rights-based approach and is quite compatible with the UNCRPD principles. It is the first law to acknowledge mental illness, such as depressive conditions, and establish a legal framework that places an emphasis on care access, anti-discrimination, and respect of autonomy of an individual. It is worth noting that the MHCA recognizes the right to mental healthcare as a justiciable right, and thus, the state has a duty of ensuring that mental health services are affordable, accessible, and of quality. Some of the MHCA provisions are specifically applicable in the case of disability on depressive illness. Rights to mental healthcare services (Section 18) would resolve one of the key obstacles to individuals affected by depression, particularly those individuals who have their functional limitations aggravated by both poverty and social marginalization. Sections on the community-based care and least restrictive treatment modalities represent the realization that institutionalization is a potential contributor of disability. Moreover, the requirements of the Act that informed consent and advance directives are taken into consideration, reflect the changing character of the depressive illness, in which the capacity may change at different times. Meanwhile, there are no challenges in the legal status of the depressive illness as a disability. Even though the MHCA lays down sound principles of normativity, the translation of these rights into practice is not even. Service providers, employers, and even the people living with depression cannot be said to be fully aware. Additionally, the overlapping of the MHCA and general disability laws in India has caused some confusion about the rights, certification, and right to social welfare benefits. In a real-life scenario, most of the depressed people still slump through these policy loopholes, especially those whose disability is less noticeable or intermittent. In theory, there is also a running debate on whether depression should be considered as a disability or not (Balakrishnan et al., 2019). There is an opinion that this labeling can lead to the reinforcement of stigma or encourage dependence. Nevertheless, practice shows that the lack of disability recognition can be even more detrimental since people do not receive accommodations, legal protection, and social support. The only way then is a balanced approach, which acknowledges disability in situations of functional impairment, but which also focuses on recovery, ability, and inclusion. Table 7.1 provides a summary of the conceptual and legal views, which are important to understand depressive illness as psychosocial disability and what it implies with regard to disability care and policy.

Altogether, the idea and legalization of the depressive illness as psychosocial disorder is a significant step to more inclusive mental health systems. Through the incorporation of the rights-based approach into the law and policy, such as the MHCA, 2017, the opportunities to respond not only to the clinical requirements but to the social implications of depression overall are presented. The task in the future is to enhance implementation and make sure that these legislative promises create positive changes in the lived lives of individuals with depressive disease.

Table 7.1 Conceptual and legal perspectives on disability in depressive illness

Dimension	Conceptual Perspective	Legal / Policy Perspective	Implications for Disability Care
Understanding of Disability	Depression-related disability viewed as functional and psychosocial impairment shaped by social context	Disability recognized through rights-based frameworks such as UNCRPD and national mental health laws	Shifts focus from symptom control to functional recovery and social inclusion
Nature of Impairment	Episodic, fluctuating, and context-dependent limitations in cognition, motivation, and social functioning	Legal recognition allows for protection even when disability is not permanent or visible	Supports flexible and individualized care pathways
Model of Disability	Transition from medical model to social and rights-based models	Emphasis on dignity, autonomy, and non-discrimination under MHCA, 2017	Encourages person-centered and least restrictive care approaches
Role of Environment	Disability emerges from interaction between individual impairment and external barriers	State obligation to reduce barriers through accessible services and supportive policies	Highlights importance of community-based and integrated services
Legal Recognition	Historically limited or ambiguous for mental illness	Explicit inclusion of mental illness under MHCA, 2017 and alignment with UNCRPD	Strengthens entitlement to care, protection, and legal safeguards
Rights and Autonomy	Capacity may fluctuate during depressive episodes	Provisions for informed consent, advance directives, and nominated representatives	Protects agency and decision-making rights of individuals
Social Participation	Depression often leads to exclusion from work, education, and community life	Legal commitment to equality and inclusion	Supports accommodations and anti-discrimination measures

7.3 Policy Frameworks Governing Disability Care in Depression

Policies are important to determine the responses of societies when dealing with disability caused by depression. Although clinical care is also essential, policies dictate the nature of people who receive services, the type of support to be given priority, and the means through which disability due to depressive illness is identified and tackled. It is apparent that mental health policy has changed over the last twenty years, becoming more inclusive and broader in treatment models with the consideration of disability, social participation, and human rights. This has been a shift that has been brought about by international commitments and also by the changing national legislations. According to the Sustainable Development Goals (SDGs) at the international level, the policy environment of disability care in depression gives a broad policy context to mental illness, although not explicitly referring to it in all details. SDG 3 that is concerned with good health and well-being highlights mental health promotion and non-communicable disease-related premature mortality reduction, including depression. Another fact that is of great significance is SDG 10, which seeks to diminish the inequalities and implicitly takes care of the social exclusion and marginalization that are frequently faced by persons with psychosocial disabilities (Tebbutt et al., 2016). Nevertheless, the translation of these general objectives into working mental health disability policies has been a thorn in the flesh of most nations. Along with the SDGs, there are also numerous international mechanisms that have played a key role in policy directions; they include the Comprehensive Mental Health Action Plan of the World Health Organization and the United Nations Convention on the Rights of Persons with Disabilities (UNCRPD). The UNCRPD, especially, has redefined the discussion where persons with mental health conditions are no longer seen as the recipients of charity or welfare. It commits signatory states to grant equality before the law, accessibility, and inclusion of persons with psychosocial disabilities in the community. In the case of depression, it implies that the policies should not be confined to hospital care but include employment, education, housing, and social participation (Nally et al., 2021). Most policy frameworks, in practice, however, continue to find it difficult to realize these commitments in specific and quantifiable forms. The policy regarding mental health and disability has developed significantly in Indian context particularly with the establishment of the Mental Healthcare Act (MHCA), 2017. The Act is a dramatic change in policy in that mental healthcare is identified as a legal right and has principles of dignity, autonomy, and non-discrimination. To persons with depressive illness, this framework can be of great help, as it provides them with valuable safeguards that include the right to affordable mental healthcare, as well as protection against inhuman or degrading treatment (Malhotra et al., 2024). MHCA also encourages home-based care and discourages unwarranted institutionalization especially when dealing with long-term disability by depression. However, with these improvements, the policy environment that regulates the situation of disability care in depression in India is disjointed. Mental health policies are usually independent of the disability welfare programs, labor policies, and social justice programs. Consequently, people who have a disability related to depression might be under clinical care without access to accommodation and income support as well as rehabilitation services at the workplace. This distinction is an expression of a more general policy trend to understand mental illness, firstly, in health terms, and not as a cross-cutting disability and development concern. Practically, it causes lapses in care continuity and constrains the usefulness of the current policies. The other critical issue is implementation and governance. Although such initiatives as the MHCA offer a powerful legal framework, policies require proper funding, human resource preparation, and coordination among the bureaucracies. Mental health services are still scarce in most of the areas and in majority of the rural and underserved regions. The developing mechanisms of monitoring and accountability arrangements are also yet to be evaluated, and it is challenging to determine whether the promises of policy are being realized in the form of enhanced functional outcomes to persons with depression. Data and outcome indicators reflecting disability specifically as opposed to clinical recovery alone are also not used extensively. To reflect the multi-layered concept of policy governance in the context of disability care when people are facing depressive illness, Fig. 7.1 draws a schematic picture of the major world, national, and sector-specific policy regimes that cumulatively define the service delivery and protection of rights.

In general, the policy frameworks that regulate disability care in depression are characterized by the process of the change toward more inclusive and rights-based approaches. The international obligations owed by countries through the SDGs and UNCRPD together with the national laws of the country (in this case, India) through the MHCA, 2017, have provided a significant base. Nevertheless, to cause meaningful change, there is a need to enhance the policy integration, introduce stronger strategies of implementation, and focus on disability outcomes better. The problem of bridging the gap between the policy intent and lived experience is one of the key challenges in the development of the disability care of depressive illness.

7.4 Policy Innovations in Disability Care for Depressive Illness

In recent times, there has been more awareness about depressive illness, which has improved the traditional ways of caring for affected people. Nowadays, there is a stronger focus on community involvement and person-centered care to help and solve this problem (Tylee et al., 1999). While these changes are currently limited and not the same everywhere, the good news is that there is an ongoing discussion on how to handle depression-related issues and complex disabilities more effectively. The shift to integrated care models has become one of the greatest policy novelties. These models are designed to fill the gap that has long existed between mental healthcare, the primary healthcare, and the welfare system. Practically, it implies the integration of depression screening and follow up into a primary care system and the connection between clinical care and rehabilitation, employment assistance, and social services (Kroenke & Unutzer, 2017). This kind of integration is especially applicable in individuals whose functional impairment is as a result of depression as the level of fragmentation and repetitive service gaps are minimized. The integration is, however, easier to facilitate at the policy level than on the ground where there is still a challenge of coordination of various departments. Another critical innovation in the disability policy on depressive illness has also led to community-based care. The focus of policies is shifting toward care provided in areas that are near residential locations instead of having only specialized facilities. In patients having a disability associated with depression, community-based models could help to alleviate the stigma and facilitate slow recovery to normal life. Nevertheless, the standard and

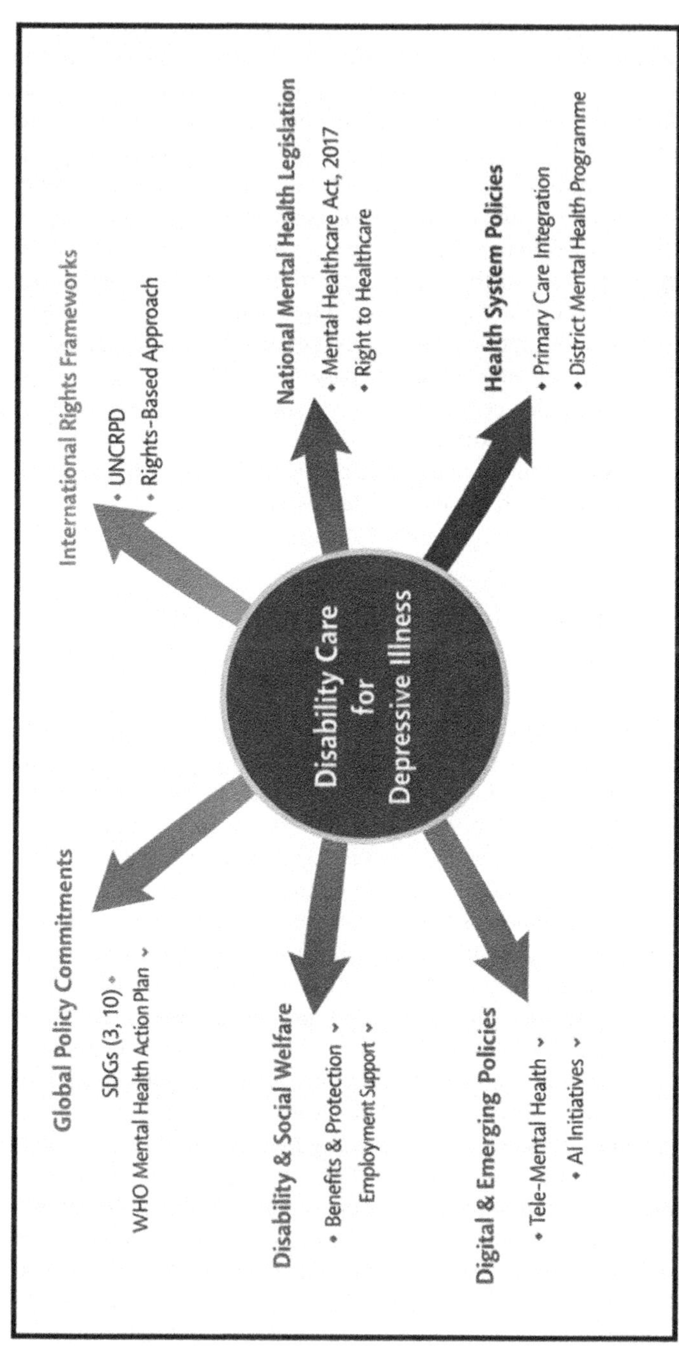

Fig. 7.1 Policy frameworks governing disability care in depression

extent of such services is highly uneven, and in most regions, they are under-staffed. The other significant field of innovation is the application of digital technologies in disability care. Due to the shortage of healthcare workers in our country and other developing nations, digital services like telepsychiatry and mobile apps have been included in policies to reach remote areas. Regarding disability, these methods are very helpful for providing continuous treatment and identifying mental health issues early (Ahmed, 2024).

Simultaneously, digital practices also bring up issues of equity, digital literacy, and data privacy, which are not effectively dealt with by the policies. Some of the edge-optimized and explainable deep learning frameworks with real-time mental health monitoring systems can enable early detection of mental disabilities while ensuring transparency in clinical decision-making to cure the depressive illness (Shukla et al., 2025). Depressive illness has also been gradual in getting work place and social protection policies innovated. The measures are particularly significant in those cases when the person experiences depression that influences concentration, the level of energy, or the quality of interpersonal functioning. Such policies can prevent temporary impairment to become permanent disability since they are aimed at retention and adaptation instead of exclusion. However, the implementation is sporadic, and most people do not know their rights or they fear getting stigmatized or lose their jobs (Secker, 2003). There is an increased motivation of persons with mental conditions to participate in service planning, implementation, and evaluation through policies. This change calls into question previous top-down strategies and acknowledges that depressed people with a disability have much to tell about what is and is not working. Although it has not yet achieved meaningful participation in most of the settings, the fact that it has come into policy-making discussions is a significant move toward more accountable and responsive systems. After these innovations, there is a need to recognize the limitations of the same. Many policy initiatives are neither long-term nor sustained. Their effectiveness is still compromised by financial limitations, manpower issues, and poor monitoring systems (Krothe, 1997). By combining new ideas (innovations) and community support, we can create interesting ways to care for people dealing with depression. Figure 7.2 shows how these different innovations work together.

7.5 Accessibility, Equity, and Inclusion Challenges

While mental health policies have improved recently, it is still a major challenge to reach and care for people with depression, especially when trying to include technology. The cause of the problem lies in social attitudes, economic inequality, and flaws in the healthcare system; therefore, we cannot treat it as just a technical or administrative failure (Packness et al., 2019).

Patients with depression often face social and administrative hurdles. Most services are only available in metro cities, making it very difficult to provide expert

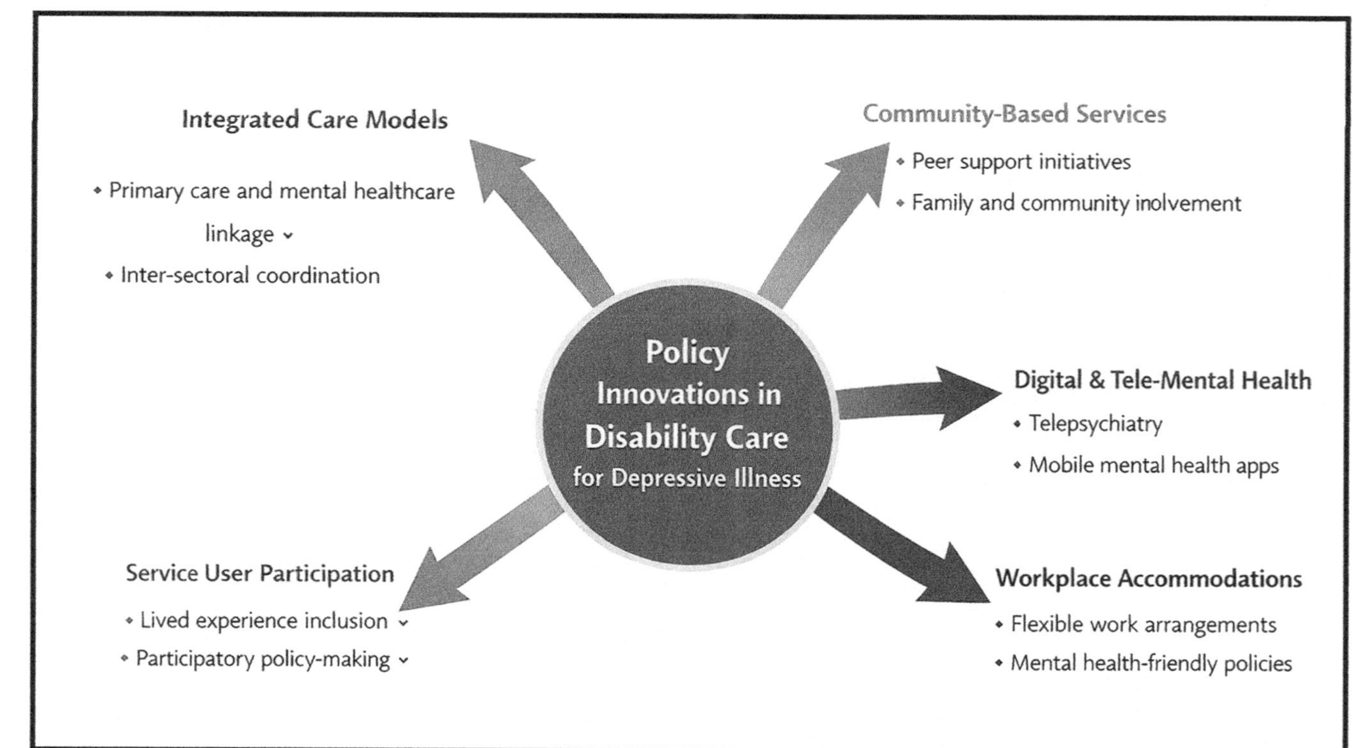

Fig. 7.2 Policy innovations in disability care for depressive illness

treatment to people living in rural or remote areas. The care can become complicated in complex referral pathways, information deficit, and limited service hours, particularly in individuals with low motivation, exhaustion, or cognitive impairments because of depression (Trujillo et al., 2025).

The issue of equity also makes access to depressive illness disability care complicated. The socioeconomic status plays a critical role in the receivers and non-receivers of care. Lower income earners are usually exposed to stressors that lead to depression but lack the benefits of seeking long-term treatments or rehabilitations. Gender can also be considered, because women can be more depressed, and at the same time, they might have an obstacle regarding the caregiving tasks, social norms, and decreased autonomy in making health-related decisions (Lemuel et al., 2021). The less privileged, such as people with disabilities, the elderly, and people in the less favored communities, are especially at risk of being forgotten in the mainstream mental health provisions. These challenges are mutually sustained as in Table 7.2, and they are multifaceted; hence, the importance of coordinated, inclusive, and rights-based approaches to care about depressive illness in the case of disability.

Integration is also a major challenge particularly in other non-clinical aspects like acquiring employment, education, and community living. The stigma of depression still has a role in the policy implementation, regardless of the legal protections in place. The employer might be unwilling to make some reasonable accommodation, and schools might not have the resources or the inclination to accommodate students with psychosocial disability (Prizeman et al., 2024). Many studies show that the reason mental health issues and depression turn into long-term disabilities lies in social and economic problems. Since these issues follow a predictable pattern, it is believed that Artificial Intelligence (AI) can help identify them by spotting early

Table 7.2 Key accessibility, equity, and inclusion challenges

Challenge	Impact on Persons with Depressive Illness	Policy Considerations
Access & Geography	Urban-centric services; long travel; delayed care	Expand community-based services; strengthen primary care
Socioeconomic Inequities	Limited ability to afford care; higher chronic disability risk	Targeted subsidies; social protection linkages
Stigma & Social Exclusion	Reluctance to seek help; social isolation	Awareness campaigns; anti-discrimination policies
Vulnerable Groups	Women, marginalized communities face additional barriers	Gender-sensitive, culturally responsive interventions
Digital & AI Barriers	Limited access to telehealth; biased algorithms	Inclusive technology design; digital literacy programs
Rights Awareness	Underutilization of entitlements; limited advocacy	Legal literacy programs; accessible grievance mechanisms

signs. Through AI technology, we can become more efficient at monitoring patients and creating treatment plans. Using technology correctly gives us the chance to adopt new approaches in mental healthcare (Levkovich, 2025).

It is true that AI algorithms are not yet 100% accurate; however, using technology and AI tools more effectively has already given us better results. At the same time, we cannot ignore the major challenges of using AI, such as social inequality, data privacy, and trust issues.

7.6 Implementation Challenges in the Indian Context

Currently, India has made significant progress in creating mental health policies and identifying depression-related illnesses. However, much more change is still needed. The Indian government introduced the Mental Healthcare Act (MHCA) 2017, which gives patients legal rights and provides a practical plan to deal with these issues. But in reality, the implementation of this law is very slow and inconsistent across the country (Pathare & Kapoor, 2020).

This disparity in policy intention and practice experience is a blend of structural, administrative, and socio-cultural forces that still determine the delivery of mental health services in the nation. Figure 7.3 gives a summary of the key implementation issues that influence disability care for depressive illness in the Indian setting, including the integration of the structural, administrative, and socio-cultural barriers.

Among the urgent issues, the lack of mental health workers deserves to be mentioned. There is an established shortage of psychiatrists, clinical psychologists, psychiatric social workers, and mental health nurses in India especially in the government healthcare systems (Sinha & Kaur, 2011). Rural and semi-urban regions are disproportionately affected by this shortage, since access to specialized care is very low in these regions. In the case of those with depression-related disability, there is not enough trained professionals and hence the delay in diagnosis, poor follow-up, and limited psychosocial rehabilitations. Despite the propagation of task-sharing and training of primary care providers as the solutions, their magnitude and regularity vary across states significantly. Another major obstacle is financing and resources allocation. The proportion of mental health in public health spending remains relatively low even though the level of depression disorders is very high. The amount of money provided by national levels fails to meet the needs of full-fledged disability care, as they need not only treatment, but also rehabilitation, social support by the community, and social protection (Bower et al., 2003). In theory, it results in excessive use of pharmacological treatment and a lack of resources to support counseling, occupational therapy, or long-term support services. The budget limitations also have influence on the development of infrastructure, especially on community-based mental healthcare services that are projected under the MHCA.

The further complication is administrative complexity and intersectoral fragmentation that make implementation more complex. Mental health and disability care is shared among various ministries and departments, such as the health, social justice,

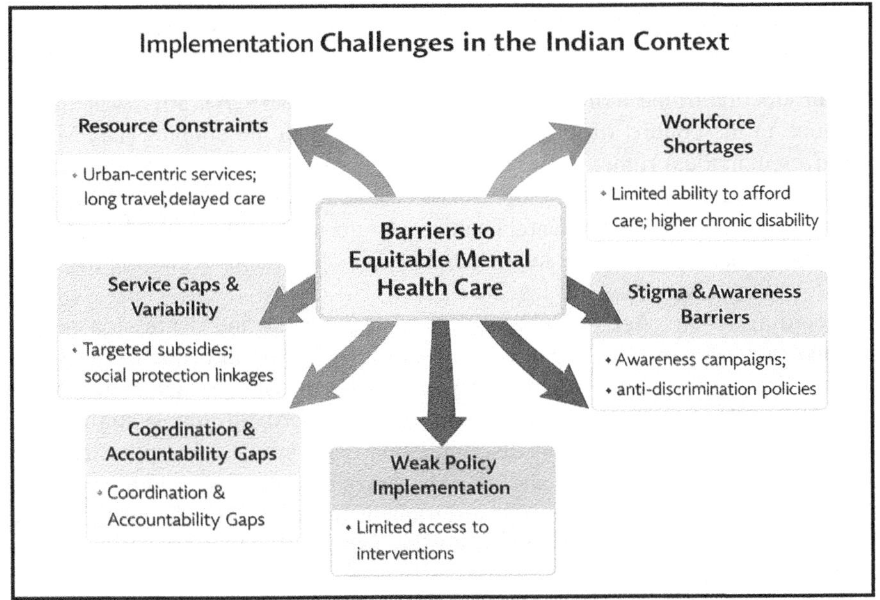

Fig. 7.3 Implementation challenges in disability care for depressive illness in the Indian context

labor, and education. Coordination among these sectors is not very strong, and it leads to disjointed service delivery (Pathare & Kapoor, 2020). At the ground level, there is still a huge lack of awareness in society. Many healthcare workers, administrators, and police officers are not yet fully aware of the rules under the MHCA, 2017.

Even the fact that depression-related illnesses are now officially recognized as a disability is not known to everyone. In most cases, depression is treated as a stigma (shame) due to deep-rooted superstitions like ghosts or evil spirits. Often, symptoms of depression are ignored or seen as normal, which delays treatment. For example, in rural areas, depression in women is frequently blamed on superstition rather than health. Additionally, factors like gender, caste, and financial status have a huge impact on mental healthcare, as they determine who gets proper treatment and who does not.

Even well-crafted policies are usually constrained by these social realities. Lastly, monitoring and evaluation systems are yet to evolve. Systematic information on the outcomes of disability with regard to depression is limited, and thus it is hard to determine the effectiveness of the policy interventions in the real world (Hyseni Duraku et al., 2024). Current indicators are more inclined to consider service usage, as opposed to functional recovery and social inclusion. In the absence of sound data, policy changes will be reactive as opposed to evidence-based. Comprehensively, the experiences of implementation in the Indian context underscore the difficulty of converting the rights-based policies on mental care into practice.

7.7 Rights-Based Mental Health, Equity, and Inclusion

The introduction of the Indian Mental Healthcare Act (MHCA), 2017 was a major milestone in the context of addressing mental health in the country. The MHCA prioritizes individual rights, dignity, and autonomy over mental healthcare, unlike previous legislation that mostly emphasized custodial care and institutional control. With this change, the Indian mental health system is now in line with global human rights standards, especially the new United Nations provisions regarding the rights of persons with disabilities (Molas, 2016).

According to this Act (MHCA, 2017), mental illness and depression are not just medical problems; they are also issues of social justice and equality. The most important part of this Act is that it clearly recognizes the right to mental healthcare. Section 18 of the Act gives every person the right to receive affordable, high-quality mental health services without any discrimination. This provision is especially meant for people from weak economic and social backgrounds who have traditionally been left out of care. Under the Act, the government must provide all levels of care, including services at primary health centers and special facilities, without any bias (Gao et al., 2022).

The Indian Mental Healthcare Act (MHCA), 2017 is a major initiative for mental health in India. This Act prioritizes a patient's personal autonomy and their consent for treatment, which is a big change from the old system where only doctors made all the decisions. It includes special provisions like 'Advance Directives' and 'Nominated Representatives.' These give patients the right to decide about their future treatment while they are in a healthy state of mind. This is especially important for depression patients, whose ability to make decisions can change from time to time.

The Act prohibits discrimination based on gender, caste, or religion and ensures that mental health services are accessible to everyone. Another important goal is to erase the social stigma attached to mental illness and help patients rejoin mainstream society.

While this Act is very strong on paper, implementing it on the ground is a major challenge. The main problems are a shortage of specialist doctors, limited resources, and the unequal distribution of services. Even today, people living in poor and remote areas lack access to good facilities. Changing society's mindset also remains a massive challenge.

In short, as shown in Table 7.3, the MHCA 2017 is a foundation for the legal framework of mental healthcare. However, its full success is only possible if the government increases investment in infrastructure, creates public awareness, and makes constant efforts to end discrimination. This Act is not just about medical treatment; it is about giving mentally ill persons the right to live with dignity.

Table 7.3 Key provisions of the Indian mental healthcare Act, 2017 and their relevance to equity and inclusion

MHCA 2017 provision	Rights-based focus	Implications for equity and inclusion
Right to Mental Healthcare	Legal entitlement to accessible and affordable care	Reduces exclusion and financial barriers
Informed Consent & Capacity	Recognition of autonomy and decision-making rights	Protects dignity and agency of persons with depression
Advance Directives & Nominated Representatives	Support for individual preferences during illness episodes	Ensures continuity and person-centered care
Least Restrictive Care	Emphasis on community-based treatment	Promotes social inclusion and reduces institutionalization
Non-discrimination	Prohibition of unequal treatment	Supports equality in health, employment, and social life
Grievance Redressal Mechanisms	Legal accountability and oversight	Enhances access to justice and protection of rights

7.8 Conclusion and Future Aspects

In this article, we discuss policy innovations and rights-based care for depression-related disability, specifically within the Indian context. According to national policies, depression is no longer seen just as a medical condition but also as a major cause of 'psychosocial disability' or depressive illness. However, from an implementation standpoint, much work remains to be done. While new innovations in Artificial Intelligence (AI) and technology are working to change the concept of mental health policy, many challenges still exist.

Regarding this issue, legal and policy changes alone are not enough. While they certainly ensure the rights, dignity, and care of a depressed person, the real problem will not end until the social stigma associated with it disappears. Many people struggling with depression still face social rejection. They also face difficulties due to a medical system that lacks adequate resources. Society should view depression not just as an illness, but in terms of a person's ability to work, participate, and their overall quality of life. Therefore, to understand the true meaning of this problem, it is necessary to adopt a disability-informed approach that is more comprehensive and realistic.

Access to care and treatment for depression remains unequal. The main reasons for this are socioeconomic inequality, gender norms, geographic location, and the digital divide. Policies that encourage community care—supported by primary health systems and social services—are better suited for depression-related disability. We can also use AI and technological innovations for the social rehabilitation of depressed individuals, and efforts are already moving in this direction.

In conclusion, the future of disability care for depression depends on rights-based legal frameworks that are inclusive, well-resourced, and accountable. In this journey, new technologies like AI, digital tools, and long-term programs will be very helpful. This requires strong political will and the active participation of patients with real-life experience. Ultimately, depression-related disability can only be tackled by adopting equality, dignity, and solid mental health policies. These efforts aim to protect human values by helping people lead meaningful and active lives within their communities.

References

Bain, P. G., Demarque, C., Park, J., Johansson, L.-O., Calligaro, C., Bushina, E., Kurz, T., Milfont, T. L., Crimston, C. R., Guan, Y., & Kroonenberg, P. M. (2019). Public views of the sustainable development goals across countries. *Nature Sustainability, 2*(9), 819–825. https://doi.org/10.1038/s41893-019-0365-4

Balakrishnan, A., Moirangthem, S., Kumar, C. N., Murthy, P., Kulkarni, K., & Math, S. B. (2019). The rights of persons with disabilities act 2016: Mental health implications. *Indian Journal of Psychological Medicine, 41*(2), 119–125. https://doi.org/10.4103/ijpsym.ijpsym_364_18

Bower, P., Rowland, N., & Hardy, R. (2003). The clinical effectiveness of counselling in primary care: A systematic review and meta-analysis. *Psychological Medicine, 33*(2), 203–215. https://doi.org/10.1017/s0033291702006979

Gao, C. X., Filia, K., Cotton, S. M., Hamilton, M. P., Rickwood, D., Simons, K., Mcdonald, L. P., Menssink, J. M., Hickie, I., Mcgorry, P. D., & Rice, S. (2022). Inequalities in access to mental health treatment by Australian youths during the COVID-19 pandemic. *Psychiatric Services, 74*(6), 581–588. https://doi.org/10.1176/appi.ps.20220345

Hyseni Duraku, Z., Davis, H., Arënliu, A., Uka, F., & Behluli, V. (2024). Overcoming mental health challenges in higher education: a narrative review. *Frontiers in Psychology*, 15. https://doi.org/10.3389/fpsyg.2024.1466060

Kemp, C. G., Concepcion, T., Ahmed, H. U., Anwar, N., Baingana, F., Bennett, I. M., Bruni, A., Chisholm, D., Dawani, H., Erazo, M., Hossain, S. W., January, J., Ladyk-Bryzghalova, A., Momotaz, H., Munongo, E., Oliveira E Souza, R., Sala, G., Schafer, A., Sukhovii, O., Collins, P. Y. et al. (2022). Baseline situational analysis in Bangladesh, Jordan, Paraguay, the Philippines, Ukraine, and Zimbabwe for the WHO Special Initiative for Mental Health: Universal Health Coverage for Mental Health. *PLoS ONE, 17*(3), e0265570. https://doi.org/10.1371/journal.pone.0265570

Kroenke, K., & Unutzer, J. (2017, April). Closing the false divide: Sustainable approaches to integrating mental health services into primary care. *Journal of General Internal Medicine, 32*(4), 404–410. https://doi.org/10.1007/s11606-016-3967-9. Epub 2017 Feb 27. PMID: 28243873; PMCID: PMC5377893.

Lemuel, A. M., Afodun, A. M., Usman, I. M., Henry, R., Yusuf, H., Kembabazi, S., Ayuba, J. T., Aigbogun, E. O., Welburn, S. C., Kairania, E., Archibong, V., Ifie, J. E., Okon, O., El-Saber Batiha, G., Assaggaf, H. M., Swase, D. T., Odoma, S., Matama, K., Ssempijja, F., … Alghamdi, S. (2021). COVID-19-Related Mental Health Burdens: Impact of Educational Level and Relationship Status Among Low-Income Earners of Western Uganda. *Frontiers in Public Health*, 9. https://doi.org/10.3389/fpubh.2021.739270

Levkovich, I. (2025). Is artificial intelligence the next co-pilot for primary care in diagnosing and recommending treatments for depression? *Medical Sciences (Basel, Switzerland), 13*(1), 8. https://doi.org/10.3390/medsci13010008

Malhotra, S., Srivastava, S., Gowda, M. R., Sharma, N., Gopalan, M. R., Watve, V. G., & Paul, I. (2024). Amend the mental health care act 2017: A survey of Indian psychiatrists (Paper 1).

Indian Journal of Psychiatry, 66(9), 829–837. https://doi.org/10.4103/indianjpsychiatry.indianjpsychiatry_667_24

Math, S., Gowda, M., Manjunatha, N., Gowda, G., Basavaraju, V., Kumar, C., Enara, A., & Thirthalli, J. (2019). Cost estimation for the implementation of the mental healthcare act 2017. *Indian Journal of Psychiatry, 61*(Suppl 4), 650. https://doi.org/10.4103/psychiatry.indianjpsychiatry_188_19

Math, S. B, Gowda, M. R., Sagar, R., Desai, N. G., & Jain, R. (2022, March). Mental health care act, 2017: How to organize the services to avoid legal complications? Indian Journal of Psychiatry, 64(Suppl 1), S16–S24. https://doi.org/10.4103/indianjpsychiatry.indianjpsychiatry_743_21. Epub 2022 Mar 22. PMID: 35599659; PMCID: PMC9122135.

Molas, A. (2016). Defending the CRPD: Dignity, flourishing, and the universal right to mental health. *The International Journal of Human Rights, 20*(8), 1264–1276. https://doi.org/10.1080/13642987.2016.1213720

Morrissey, F. (2012). The United nations convention on the rights of persons with disabilities: A new approach to decision-making in mental health law. *European Journal of Health Law, 19*(5), 423–440. https://doi.org/10.1163/15718093-12341237

Nally, D., Moore, S. S., & Gowran, R. J. (2021). How governments manage personal assistance schemes in response to the United Nations Convention on the Rights of Persons with Disabilities: A Scoping Review. *Disability & Society, ahead-of-print (ahead-of-print)*, 1728–1751. https://doi.org/10.1080/09687599.2021.1877114

Packness, A., Waldorff, F. B., Halling, A., Hastrup, L. H., & Simonsen, E. (2019). Are perceived barriers to accessing mental healthcare associated with socioeconomic position among individuals with symptoms of depression? Questionnaire-results from the Lolland-Falster Health Study, a rural Danish population study. *British Medical Journal Open, 9*(3), Article e023844. https://doi.org/10.1136/bmjopen-2018-023844

Pathare, S., & Kapoor, A. (2020). Implementation update on mental healthcare act, 2017 (pp. 251–265). Springer. https://doi.org/10.1007/978-981-15-5009-6_11

Prizeman, K., Weinstein, N., & Mccabe, C. (2024). Strategies to overcome mental health stigma: Insights and recommendations from young people with major depressive disorder (MDD). *Brain and Behavior, 14*(9). https://doi.org/10.1002/brb3.70028

Sánchez, J., Umucu, E., Iwanaga, K., Chan, F., Crespo-Jones, M., Brooks, J. M., Muller, V., & Tu, W.-M. (2018). Personal and environmental contextual factors as mediators between functional disability and quality of life in adults with serious mental illness: A cross-sectional analysis. *Quality of Life Research, 28*(2), 441–450. https://doi.org/10.1007/s11136-018-2006-1

Secker, J., & Membrey, H. (2003, April). Promoting mental health through employment and developing healthy workplaces: The potential of natural supports at work. *Health Education Research, 18*(2), 207–215. https://doi.org/10.1093/her/18.2.207. PMID: 12729179.

Shukla, A., Khatana, K., Nithya, P., Shankar, B., Vankayalapati, R. K., & Dubey, S. (2025). Edge-optimized and explainable deep learning framework for real-time intrusion detection in industrial Iot. *SSRN Electronic Journal.* https://doi.org/10.2139/ssrn.5077557

Sinha, S. K., & Kaur, J. (2011). National mental health programme: Manpower development scheme of eleventh five-year plan. *Indian Journal of Psychiatry, 53*(3), 261. https://doi.org/10.4103/0019-5545.86821

Tebbutt, E., Maclachlan, M., Horvath, R., Khasnabis, C., Borg, J., & Brodmann, R. (2016). Assistive products and the Sustainable Development Goals (SDGs). *Globalization and Health, 12*(1). https://doi.org/10.1186/s12992-016-0220-6

Trujillo, L. M. G., Cruz, L. K. G., Laguado, J. S. C., & Rincón, E. H. H. (2025). Barriers to accessing pediatric healthcare in Latin America: A scoping review. *Journal of Racial and Ethnic Health Disparities.* https://doi.org/10.1007/s40615-025-02510-w

The manufacturer's authorised representative in the EU is Springer Nature Customer Service Centre GmbH, Europaplatz 3, 69115 Heidelberg, Germany. If you have any concerns regarding our products, please contact ProductSafety@springernature.com

Printed and bound by CPI Group (UK) Ltd, Croydon, CR0 4YY

07/07/2026

02160931-0005

Creating Brand in the Building Automation, Controls, and Smart Buildings Space

Through a Look at the Niagara Framework®

By Marc Petock

For permission requests, contact:
Logicmaker Intelligence, LLC
jacob@logicmakerintel.com

Trademark Notice
All product names, logos, brands, and trademarks mentioned in this book are the property of their respective owners. Use of these names is for identification purposes only and does not imply endorsement.

Niagara Framework® is a registered trademark of Tridium, Inc.

Disclaimer
The information in this book is provided for educational and informational purposes only. While efforts have been made to ensure accuracy, the author and publisher make no warranties regarding completeness or applicability and assume no liability

for damages resulting from the use of this information. Readers should consult qualified professionals before implementing any strategies discussed.

ISBN: 979-8-9988130-3-0
First Edition: May 2026
Printed in the United States of America

LOGICMAKER INTELLIGENCE, LLC
Huntersville, North Carolina
www.logicmakerintel.com

Independent publishing focused on building automation, controls, and smart infrastructure.